Peter Schüpbach

Influence of Purge Flow on Performance of an End Wall Profiled Turbine

Peter Schüpbach

Influence of Purge Flow on Performance of an End Wall Profiled Turbine

Südwestdeutscher Verlag für Hochschulschriften

Impressum/Imprint (nur für Deutschland/ only for Germany)
Bibliografische Information der Deutschen Nationalbibliothek: Die Deutsche Nationalbibliothek verzeichnet diese Publikation in der Deutschen Nationalbibliografie; detaillierte bibliografische Daten sind im Internet über http://dnb.d-nb.de abrufbar.

Alle in diesem Buch genannten Marken und Produktnamen unterliegen warenzeichen-, marken- oder patentrechtlichem Schutz bzw. sind Warenzeichen oder eingetragene Warenzeichen der jeweiligen Inhaber. Die Wiedergabe von Marken, Produktnamen, Gebrauchsnamen, Handelsnamen, Warenbezeichnungen u.s.w. in diesem Werk berechtigt auch ohne besondere Kennzeichnung nicht zu der Annahme, dass solche Namen im Sinne der Warenzeichen- und Markenschutzgesetzgebung als frei zu betrachten wären und daher von jedermann benutzt werden dürften.

Verlag: Südwestdeutscher Verlag für Hochschulschriften Aktiengesellschaft & Co. KG
Dudweiler Landstr. 99, 66123 Saarbrücken, Deutschland
Telefon +49 681 37 20 271-1, Telefax +49 681 37 20 271-0
Email: info@svh-verlag.de
Zugl.: Zürich, ETH, Diss., 2009

Herstellung in Deutschland:
Schaltungsdienst Lange o.H.G., Berlin
Books on Demand GmbH, Norderstedt
Reha GmbH, Saarbrücken
Amazon Distribution GmbH, Leipzig
ISBN: 978-3-8381-1467-5

Imprint (only for USA, GB)
Bibliographic information published by the Deutsche Nationalbibliothek: The Deutsche Nationalbibliothek lists this publication in the Deutsche Nationalbibliografie; detailed bibliographic data are available in the Internet at http://dnb.d-nb.de.

Any brand names and product names mentioned in this book are subject to trademark, brand or patent protection and are trademarks or registered trademarks of their respective holders. The use of brand names, product names, common names, trade names, product descriptions etc. even without a particular marking in this works is in no way to be construed to mean that such names may be regarded as unrestricted in respect of trademark and brand protection legislation and could thus be used by anyone.

Publisher: Südwestdeutscher Verlag für Hochschulschriften Aktiengesellschaft & Co. KG
Dudweiler Landstr. 99, 66123 Saarbrücken, Germany
Phone +49 681 37 20 271-1, Fax +49 681 37 20 271-0
Email: info@svh-verlag.de

Printed in the U.S.A.
Printed in the U.K. by (see last page)
ISBN: 978-3-8381-1467-5

Copyright © 2010 by the author and Südwestdeutscher Verlag für Hochschulschriften Aktiengesellschaft & Co. KG and licensors
All rights reserved. Saarbrücken 2010

Contents

1 Introduction   **1**
- 1.1 Literature Review 3
  - 1.1.1 Secondary Flows 3
  - 1.1.2 Tip Leakage 8
  - 1.1.3 Unsteady Flow Interaction 8
  - 1.1.4 Rim Seal Purge Flow 12
  - 1.1.5 Losses 13
  - 1.1.6 Secondary Flow Control 15
  - 1.1.7 Non-Axisymmetric End Wall Contouring 17
- 1.2 Research Objectives 17
- 1.3 Thesis Outline 18

2 Research Facility and Methods   **21**
- 2.1 Experimental Facility 21
  - 2.1.1 Overview 21
  - 2.1.2 1.5-Stage Turbine 22
  - 2.1.3 Bypass and Flow Injection System 23
- 2.2 Data Acquisition 25
  - 2.2.1 Operating Conditions 25
  - 2.2.2 Vibration 27
  - 2.2.3 Multi-Channel Pressure Measurement System 27
- 2.3 Vane Instrumentation 28
  - 2.3.1 Vane Surface Pressure Measurements 28
  - 2.3.2 Vane End Wall Pressure Measurements 28
  - 2.3.3 Vane Intra-Row Traversing 28
- 2.4 Probe Technology 30
  - 2.4.1 Introduction 30
  - 2.4.2 Calibration 32
  - 2.4.3 Traversing System and Resolution 34
- 2.5 Data Reduction 35
  - 2.5.1 Pneumatic Probe Data 35

	2.5.2	FRAP 2-sensor	37
	2.5.3	Secondary Kinetic Energy	38
	2.5.4	Streamwise Vorticity	40
2.6		Uncertainty Analysis	43
2.7		CFD	48
	2.7.1	Steady CFD	48
	2.7.2	Time-Resolved CFD	50

3 Non-Axisymmetric End Wall Profiling — 55

3.1		Design Methodology	55
3.2		Operating and Inlet Conditions of the Turbine	58
3.3		Total-to-Total Stage Efficiency	61
3.4		First Vane	61
	3.4.1	End Wall Pressure Field	61
	3.4.2	Lift Plots	62
	3.4.3	Vane Suction Side Pressure Field (CFD)	65
	3.4.4	First Vane Intrarow Measurement	67
	3.4.5	First Vane Row Exit Flowfield	69
3.5		Rotor Exit Flowfield	78
3.6		1D Effects of the End Wall Contouring	82
	3.6.1	Change in Reaction	82
	3.6.2	Secondary Kinetic Energy & Loss	83
3.7		Time-Resolved Flow Physics	86
3.8		Summary	96

4 Purge Flow Effects — 99

4.1		Baseline Turbine Purge Sensitivity	99
4.2		Measured Purge Flow Effect at Rotor Inlet	100
4.3		Measured Purge Flow Effect at Rotor Exit	101
	4.3.1	Time-Averaged	101
	4.3.2	Time-Resolved	105
	4.3.3	Spectral Analysis	111
4.4		Time-Resolved Calculation	113
	4.4.1	Static Pressure at the Rim Seal Exit	113
	4.4.2	Purge Flow Mechanism	114
4.5		Summary	121

5 Influence of Purge Flow on End Wall Profiling — 123

5.1	Measured Efficiency Sensitivity to Purge	123

5.2		Non-Axisymmetric End Walls with $IR = 0.9\%$	127
	5.2.1	Influence of Vane Hub Profiling on Static Pressure at the Rim Seal Exit	127
	5.2.2	Non-Dimensional Rim Seal Exit Pressure Field	127
	5.2.3	Purge Flow Structure	130
	5.2.4	Measured Rotor Exit Flowfield	133
	5.2.5	Spectral Analysis of Measurements	135
5.3		Efficiency Sensitivity Kink	137
	5.3.1	Measured Rotor Exit Flowfield	138
	5.3.2	Spectral Analysis	144
	5.3.3	Pressure Field Sensitivity to Increasing Purge	145
	5.3.4	Jet Structure and Vorticity Development with Increasing Purge	148
5.4		Summary	151

6 Conclusion and Contribution   155
6.1 Suggestion for Further Work 159

A Nomenclature   161
A.1 Symbols 161
A.2 Greek 162
A.3 Subscripts 162
A.4 Abbreviation 164

B List of Publications   165

1 Introduction

With rising oil prizes fuel becomes the dominant cost factor in all airlines and demands all their focus. Smith [?] showed that the fuel costs are on average 16% of the direct operational costs at a crude oil price of 24$ the barrel. If the crude oil price reaches 50$ the fuel costs already reach 25% of the direct operational costs. At oil prices above 100$ a barrel the fuel cost share is even higher. This forces the airline companies to invest into new fuel saving airframes. On the other hand global warming is one of the main challenges humanity is facing within the next decades and therefore, it ranks at the top of the political agenda. From the year 2003 to the year 2006 the average number of news paper articles on global warming increased by a factor of 4 [?]. The only real countermeasure is a profound cut down in carbon emissions. At the same time the airline market in China is growing at high speed with about 8.6% per year [?].
The main potential in fuel reduction lies in the engines themselves. As stated by Waitz [?] 57% of the possible aircraft energy consumption can be achieved due to increases in engine efficiencies. The measure of engine efficiency is the specific fuel consumption (SFC), which is the fuel massflow per jet engine thrust output. Wisler [?] showed that in a high-pressure turbine 1% aerodynamic efficiency improvement leads to 0.82% improvement in SFC. In a low pressure-turbine the relation is close to equality. With respect to these circumstances loss reductions in the turbine and with this a reduction in specific fuel consumption are of high economical and social interest.
The efficiency of the engines can be changed from two sides. Either the component efficiencies can be increased or the overall engine performance by increasing the thermal efficiency. The thermal efficiency can be increased either by an elevated turbine inlet temperature or by a higher compression ratio. With higher compression ratios and predominantly with higher inlet turbine temperatures thermal management becomes an issue. However, the higher the cooling massflows are the lower the engine efficiencies become. Furthermore, in aircraft engines not only the efficiency is of importance but

also the weight of the engine matters. Therefore, a reduction of the number of components is attractive. Reducing the number of components at the same time reduces the manufacturing and maintenance costs.

Considering the above mentioned motivations the driving research objectives in aero-engine axial turbines are either improvements in thermal management or an increase in the aerodynamic efficiency. With a reduction of the number of aerodynamic components the goal is not only an increase of efficiency but rather to keep the same aerodynamic efficiency. Because if the number of stages and blades per row are reduced the stage loading and lift coefficients are continuously pushed up. As a consequence the secondary losses rise and become the dominant loss source leading to a deterioration of the efficiency. They can reach up to half of the total loss (Dunham et al., [?]). In low aspect ratio turbines the losses are even more pronounced as the hub and tip secondary flow structures start to interact with each other. Therefore, methods to reduce secondary losses are of great value to increase or at least maintain the aerodynamic efficiency.

One very promising technique is non-axisymmetric end wall profiling. Although state of the art turbine stage efficiencies range up to more than 90%, there is still room for improvement. Harvey et al. [?] have demonstrated a 0.9% ± 0.4% improvement for the intermediate turbine stage of the Rolls-Royce trent 500 engine. However, apart of these Rolls-Royce rig tests with relatively simple instrumentation no rotating rig investigations on non-axisymmetric with non-axisymmetric end walls have been published. This provides part of the motivation for the current study.

At the same time secondary cooling air is indispensable for thermal management of the aerodynamic components, because the actual turbine inlet temperatures are well beyond the melting temperature of the components. One example of secondary cooling air is purge flow which is used to prevent ingestion at the stator-rotor rim seals. Ong et al. [?] among others have shown that purge flow has a profound effect on the secondary flow structure development downstream of the rim seal. Therefore, these secondary cooling air effects should be included in investigations, which deal with secondary flow control techniques as non-axisymmetric end wall profiling .

This work presents the first combined experimental investigation of purge flow effects and non-axisymmetric end wall profiling in a rotating rig with high accuracy time-resolved instrumentation.

1.1 Literature Review

1.1.1 Secondary Flows

As mentioned in the introduction secondary losses are one of the major loss sources in axial turbines. The secondary losses are generated when secondary flows are mixing out. Secondary flow itself is an artificial quantity. Cumpsty [?] described it as the flow at right angles to the intended primary flow. Secondary flows normally are the result of streamwise vorticity. The classical secondary theory was originally proposed by Squire and Winter [?]. This inviscid theory states, that the normal vorticity introduced by the inlet boundary layer is transformed into a normal and a streamwise vortex component as result of the turning inside an airfoil cascade. It was stated that the streamwise vorticity ω_s at cascade exit is twice the normal vorticity ω_n contained in the inlet boundary layer times the turning angle of the flow as seen in Equation 1.1.

$$\omega_{s2} = -2\omega_{n1}(\alpha_2 - \alpha_1) \tag{1.1}$$

Figures 1.1 and 1.2 show the cascade vorticity as predicted by the classical secondary flow theory.

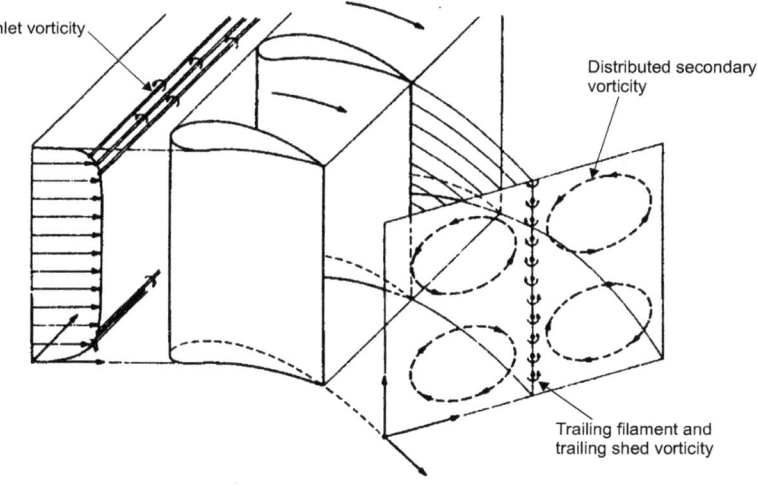

Figure 1.1: Classical Secondary Flow Model by Hawthorne [?]

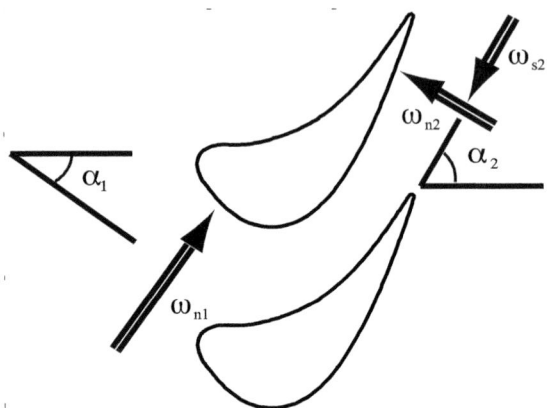

Figure 1.2: Classical Secondary Flow Model

Sieverding [?] gives a very detailed review of secondary flow investigations. A more recent review is presented by Langston [?]. These two review papers present a number of experimental investigations in linear and annular cascades. These revealed that secondary flows are highly three-dimensional and that viscous effects play an important role. Sharma and Butler [1987] from a detailed study of the available experimental data came to the conclusion that the effect of the inlet boundary layer as predicted by the classical secondary flow theories is incorrect for turbomachinery configurations. They state that the formation of the leading edge horseshoe vortex additionally transforms normal vorticity into streamwise vorticity independent of flow turning. There are typical secondary flow features independent of the individual turbine geometry. Two representative models can be seen in Figures 1.3 and 1.4.

Horseshoe Vortex The interaction of the inlet boundary layer with the blade leading edge results in a three-dimensional separation seen as a saddle point in flow visualization experiments (Langston et al. [?]). In between the separation saddle point and the blade leading edge a vortex forms. This vortex is referred to as horseshoe vortex. The horseshoe vortex consists of a pressure and suction side limb. Due to the separation process the normal vorticity of the inlet boundary layer is converted into streamwise vorticity. The two limbs of the horseshoe vortex are of opposite rotation.

1.1 Literature Review

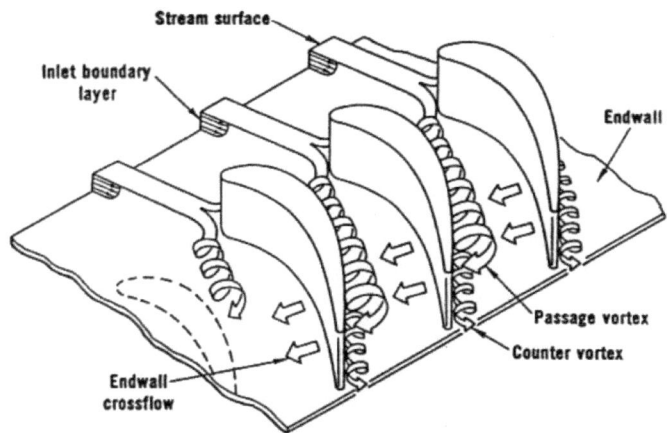

Figure 1.3: Secondary Flow Model of Langston [?]

The pressure side limb is convected across the passage as result of the cross passage pressure gradient. The point where the pressure side limb impinges onto the blade suction side is normally defined as the starting point of the passage vortex. The suction side leg of the horseshoe vortex follows the suction side. Once the pressure side limb meets the suction side the suction side limb is forced to lift off the end wall. Sieverding and Van den Bosche [?] stated that the position of the suction side limb depends strongly on the rotational speed of the passage vortex which depends on the turning. Gregory-Smith et al. [?] observed that the suction side limb of the horseshoe vortex is mixing out within the passage.

Passage Vortex Behind the separation line of the inlet boundary layer a new boundary layer is developing. The pressure side limb of the horseshoe vortex entrains this low-momentum fluid and subsequently forms the passage vortex. As stated by Sharma and Butler [?] this entrainment strongly influences the development of the passage vortex and is a key mechanism in the generation of secondary flows and end wall losses. At the same time the entrainment strongly depends on the strength of late passage streamwise perpendicular pressure gradient, which is a result of the blade turning. The passage vortex is the most prominent flow structure at the exit of shrouded turbine blade rows.

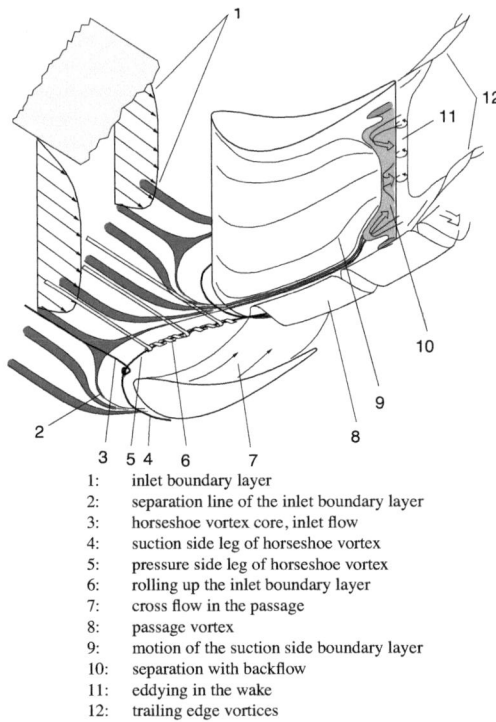

Figure 1.4: Secondary Flow Model of Vogt and Zippel [?]

1: inlet boundary layer
2: separation line of the inlet boundary layer
3: horseshoe vortex core, inlet flow
4: suction side leg of horseshoe vortex
5: pressure side leg of horseshoe vortex
6: rolling up the inlet boundary layer
7: cross flow in the passage
8: passage vortex
9: motion of the suction side boundary layer
10: separation with backflow
11: eddying in the wake
12: trailing edge vortices

Corner Vortex The corner vortex as described by Sieverding [?] is a vortical structure rotating in the opposite direction of the passage vortex and located in the end wall suction side corner. Its formed when a strong crossflow on the end wall due to high turning meets the suction side. This feature is relatively small and therefore, rarely observed in experimental investigations. Yamamoto [?, ?] presented results for two cascades with different turning. In the case with 110° turning representing a typical rotor cascade the effect of the corner vortex is seen as the result of reduced underturning close to the end wall. The other case representing a stator cascade with turning of 68° didn't show this effect. This lead to the conclusion that a corner vortex only develops beyond a certain level of turning.

1.1 Literature Review

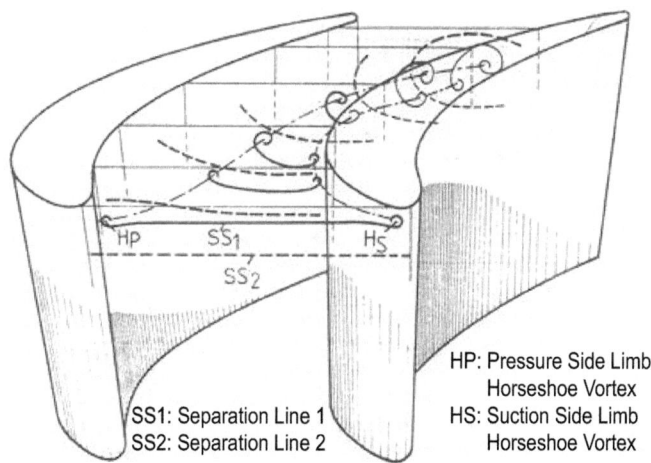

Figure 1.5: Sieverding and Van den Bosche Secondary Flow Model [?]

The influence of the incoming boundary layer thicknes on the development of the secondary flow is presented in the work of Gregory-Smith et al. [?]. They concluded that the intensity of the secondary flow structures increases with an increasing boundary layer thickness, while the position of the structures isn't affected. Sharma and Butler [?] stated from investigations with different inlet boundary layer thicknesses that the boundary loss convects through the passage without causing additional loss.

Langston [?] concluded in his review that accurate routine predictions of secondary losses have not yet been achieved. The main reasons for this are inadequate turbulence models and a still limited knowledge of end wall loss production mechanisms.

The main differences regarding secondary flows between rotating machines and linear cascades apart of the unsteady effects, which will be addressed in subsection 1.1.3, are the annular shape which introduces radial pressure gradients and the strong skewing of the inlet boundary layer caused by the relative movement of stationary and rotating rows. Walsh [?] investigated the effect of inlet skew in turbines and concluded that the situation gets worse in terms of loss. Moustapha [?] compared an annular with an linear cascade and found higher and more concentrated losses in the annular cascade. Furthermore, real machines often need cooling in order to protect the material. The cooling injection strongly affects the secondary flows. In subsection 1.1.4 rim seal purge flow as one example of a secondary cooling flow is addressed.

1.1.2 Tip Leakage

If blade rows are unshrouded an additional flow feature is introduced. The tip leakage vortex which is the result of an over the blade tip flow from the pressure to suction side. This leakage flow potentially rolls up into a vortex causing considerable aerodynamic losses. Booth [?] found the losses to be in the order of up to one third of the overall losses. Denton [?] concluded that the tip leakage loss is proportional to the discharge coefficient of the tip gap. Furthermore he concluded that the losses scale with the tip loading of the blade. Behr et al. [?] have presented a novel approach for controlling the rotor tip leakage vortex by injecting cooling air from the stationary casing onto the rotor tip. They observed a reduction in size and turbulence intensity of the tip leakage vortex as well as of the tip passage vortex. With the appropriate injection rate and axial injection position the isentropic stage efficiency was increased by 0.55%.

1.1.3 Unsteady Flow Interaction

Due to the relative motion of stationary and rotating rows an unsteady flow field is generated. The unsteadiness is normally divided into two groups:

- a random part which is the result of turbulence

- a periodic part which is related to the blade passing frequency

In order to quantify the importance of the periodic unsteadiness the non-dimensional quantity reduced frequency $\overline{f}$ is often used. It is the ratio of the convection time t_{conv} and the inverse of the disturbance frequency f_{dist}.

$$\overline{f} = \frac{t_{conv}}{1/f_{dist}} \qquad (1.2)$$

When $f \ll 1$ then the problem is quasi-steady. At $f \sim 1$, unsteady and quasi-steady effects are of the same order. If $f \gg 1$ the unsteady effects dominate.

1.1 Literature Review

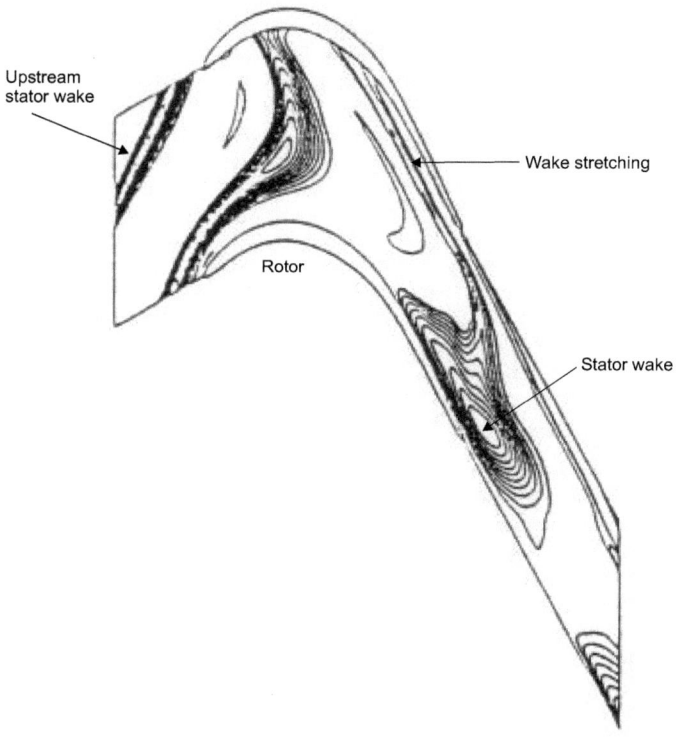

Figure 1.6: Vane Wake Deformation [?]

As the result of relative motion of rotating and stationary rows the flow features of the upstream row are alternatively facing an open channel or the leading edge of the subsequent row. Sharma and Butler [?] separated the flow downstream of the rotor into two different events. At maximum interaction the vane flow structures interact with the leading edge. Finally they end up in the rotor wake or secondary flow structures. As a result there is a very low turbulence level in the rotor free stream. At minimal interaction the vane flow structures enter the rotor passage without interaction with the rotor blades. Therefore, the free stream turbulence level is much more elevated. In order to define the interaction behaviour Schlienger [?] has defined a relative throat which is the downstream area the flow faces dependent on the relative vane blade position.

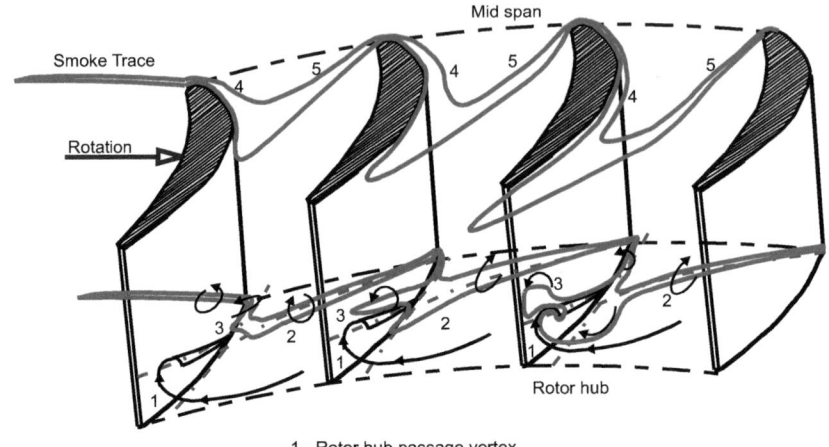

1 - Rotor hub passage vortex
2 - Pressure side leg of stator hub passage vortex
3 - Suction side leg of stator hub passage vortex
4 - Suction side leg of stator wake
5 - Pressure side leg of stator wake

Figure 1.7: Vortex Blade Interaction [?]

Potential Flow Interaction Each row generates a pressure field. The pressure field is a potential field which acts in contrast to the convective mechanisms both in up- and downstream direction. As a result of the interaction of vanes and blades a time-varing pressure field is generated. The temporal variation of pressure is a work process as stated by Dean [?]. As a consequence the upstream potential field effect generates variations in the total pressure field, often called bow waves. As stated by Parker and Watson [?] the potential effect is decaying exponentially with distance.

Wake-Blade Interaction Behind the trailing edge of an airfoil the suction and pressure side boundary layers merge and form a velocity deficit, also known as the wake. Due to the relative motion of vanes and blades the wakes are cut into lumps of low momentum fluid by the subsequent blade row. This wake-blade interaction was first investigated by Meyer [?]. Meyer introduced the so called 'negative jet' model, which is essentially a perturbation of the uniform flow. The consequence of the 'negative jet' effect is a build up of wake fluid on the blade suction side and a removal on the pressure side. As described by Hodson and Dawes [?] the effect can be

1.1 Literature Review

summarized as follows. The wake undergoes 'bowing' before it enters the blade passage due to higher velocities at the passage center. At the suction side the wake fluid is sheared due to higher convection velocities. At the pressure side the wake experiences stretching. As a result most of the wake fluid ends up at the suction side with a tail reaching over to the pressure side. Figure 1.6 shows the process of vane wakes convecting through the downstream blade row as calculated by Hodson and Dawes [?]. Hodson and Dawes [?] have also shown that the wake as it impinges onto the suction side is creating a recirculation. This recirculation creates variations in pressure. As a consequence of the convection of the wake fluid, temporal gradients of pressure are created. As stated by Dean [?] temporal gradients of pressure are a work mechanism. This results in fluctuations of stagnation temperature which are much more significant than the defects due to the wakes at blade row inlet. Rose and Harvey [?] argued that the free-stream work and wake work are not the same. They introduced the concept of "Differential Work". In a turbine less work is extracted from the wake than from the free-stream fluid. As a consequence mixing losses are reduced. Their model states that the higher the wake total temperature and pressures deficit is the lower the work extraction from the wake is compared to the free-stream.

Vortex Interaction In low aspect ratio turbines, the secondary flow structures are occupying a large portion of the passage. Binder et al. [?] stated that the vane passage vortices are cut off by the blades. They assumed that the vortex is breaking down during this process. As a result of this the kinetic energy is converted into turbulence. In contrast Chaluvadi [?] and Behr [?] have shown that the vorticies are not chopped but bent around the blade leading edge. As a result the upstream vane vortex is forming a suction and pressure side limb. Chaluvadi [?], using smoke traces, identified the suction side leg of the upstream vane passage vortex above the newly developed passage vortex. It is stated that the pressure side limb of the upstream passage vortex is merging with the forming blade passage vortex. An illustration of his model is shown in Figure 1.7. Behr [?] on the other hand identified the suction side leg of the upstream vane passage vortex below the newly developed blade passage vortex.

1.1.4 Rim Seal Purge Flow

In order to improve the thermal cycle efficiency of gas turbines, turbine entry temperatures have been continuously increased over the past decades. With these increases the ingestion of hot gases into the disk cavities has become an issue, as it can cause overheating of the disks as well as thermal fatigue of the components. In order to mitigate the adverse effects of ingestion of hot gases, bypassed compressor air is used as purge flow and injected through the rim seals between the rotating and stationary parts. Figure 1.8 shows both the ingestion as well as the purge flow.

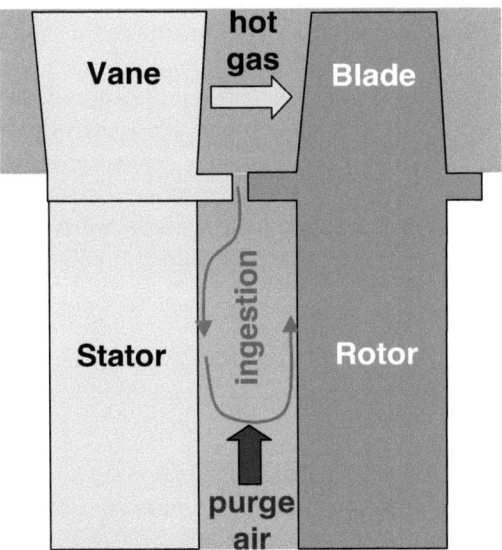

Figure 1.8: Rim-Seal Flow Sketch [?]

The goal is to minimize the amount of purge massflow and to reduce the aerodynamic losses, which can be attributed to the purge flow. The ingestion of hot gases is driven by both disk pumping as well as by the external non-axisymmetric pressure field. This has been experimentally investigated in previous studies; Kobayashi et al. [?] found that the pressure difference criterion underestimates the minimum required cooling flow rate. Chew et al. [?] and Dadkhah et al. [?] as well examined the question of the minimum coolant flow that is required to prevent ingestion. Furthermore

1.1 Literature Review

did they show where the ingested air is ending up in the cavity. However, the pressure field at rim seal exit is highly unsteady due to stator-rotor interactions. Roy et al. [?] for example showed that the effect of the unsteady pressure field was much more pronounced inside the cavity than the time-averaged circumferential external pressure field. Recent research has now as well focused on the flow interactions between the purge flow and the mainstream flow. McLean et al. [?, ?] compared the effect of radial, impingement and root injection cooling configurations. Figure 1.9 illustrates these three different cooling strategies. The root injection showed an average total-to-total efficiency increase of 1.5% They also found that the root injection caused a reduction of the under- and overtuning. Girgis et al. [?] compared radial injection to compound injection, which has both a radial and tangential component. They observed that the latter resulted in an efficiency improvement. Ong et al. [?] also concluded that some of the efficiency penalty due to coolant could be regained by introducing a swirl component to the coolant jet. Furthermore they found that most of the coolant is entrained by the downstream blade hub secondary flow. Paniagua et al. [?] reported that there is an intensification of the rotor hub vortex and an enhancement of the radial migration due to injection. In recent studies the importance of the unsteady interaction of the freestream and the cavity were highlighted. Boudet et al. [?] found frequencies that are unrelated to the blade passing frequency. They attributed this to a non-linear coupling of the blade passing frequency with an instability formed inside the cavity. They concluded that only full annulus and unsteady modelling would capture the experimentally observed flow phenomena. Reid et al. [?] quantified the efficiency penalty for a low speed machine caused by the rim seal flow to be about 0.56% for 1.0% of injection massflow. In a numerical study Marini et al. [?] examined the effects of the blade leading edge platform and noted that there is a 0.07% stage efficiency benefit and a reduced sensitivity to an increasing cavity massflow with an appropriate platform design.

1.1.5 Losses

Although the losses are often divided into three categories referred to as 'profile loss', 'end wall loss' or 'leakage loss', they are rarely independent from each other. Denton [?] introduced entropy as the only reliable measure of loss in turbomachines. The advantage of entropy is that it is independent

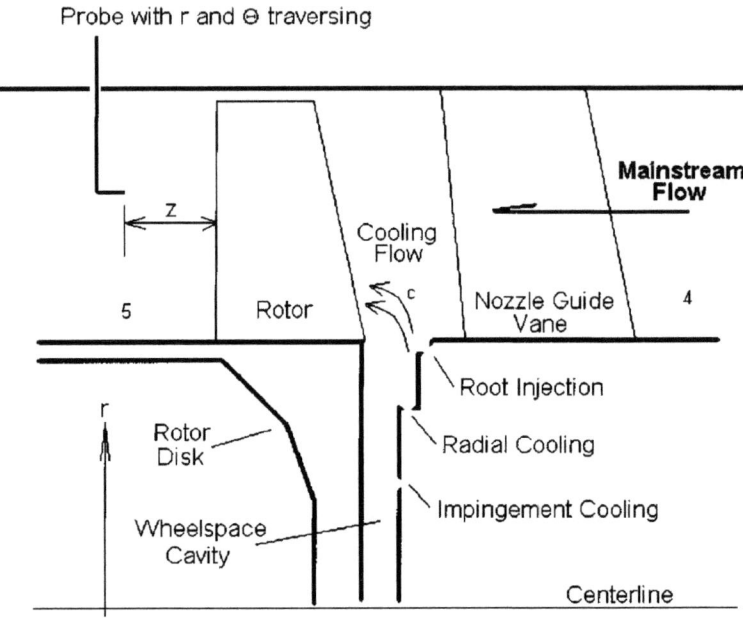

Figure 1.9: Illustration of the Different Cooling Strategies Investigated by McLean et al. [?]

of the frame of reference. However, entropy can't be measured directly and has to be calculated with pressures and temperatures.

$$s - s_{ref} = c_p \ln\left(T/T_{ref}\right) - R \ln\left(p/p_{ref}\right) \qquad (1.3)$$

Denton [?] identified three processes which create entropy:

- *Viscous Friction*
- *Heat Transfer*
- *Non-Equillibrium Processes*

Furthermore Denton introduces the term end wall loss instead of secondary loss. He calls it "*the most difficult loss component to understand*". The total entropy generation in the end wall boundary layers upstream of, within, and

1.1 Literature Review

downstream of the blade row explains 2/3 of the end wall loss. 1/4 of the total loss is associated with secondary flows.
As already mentioned in subsection 1.1.1 secondary flow doesn't introduce loss directly. However, due to high velocity gradients associated with secondary flows viscous dissipation is converting the secondary kinetic energy into losses.

1.1.6 Secondary Flow Control

Generally secondary flow control can be divided into active and passive methods. Examples of active methods are boundary layer blowing investigated by Sturm et al. [?] and Biesinger [?]. Biesinger injected air tangentially through an upstream slot to oppose the secondary flow production. However, no net loss reduction was achieved. However, some blowing configuration lead to a reduction in mixed out losses.
However, most studies have reported on passive methods. The most frequently used are blade leaning, axisymmetric and non-axisymmetric end wall profiling. Harrison [?] investigated in detail the effect of blade leaning. He reported that there is no net loss reduction within the row in which blade leaning is applied. The reason for this is that the losses at the end wall are reduced at the expense of higher mid-height losses. The benefit with blade leaning comes rather from a more homogenous flowfield going into the subsequent blade row.
The concept of axisymmetric end wall profiling was introduced by the russian engineer Dejc [?] as a contraction of the annulus from the leading edge to the trailing edge. A loss reduction of up to 20% was verified in linear cascade tests by Morris et al. [?]. The most promising end wall geometry incorporated a strong contraction early in the passage (often known as the "Russian kink"). The acceleration results in a thinner boundary layer. Atkins [?] investigated different end wall contours in a linear cascade. He showed that the losses near the end wall can be influenced by the shape and the resulting pressure field. Dossena et al. [?] reported a reduction of 35% for the overall loss and a reduction of 54% for the secondary loss with axisymmetric profiling. However, it is suspected that the reduction is the result of reduced loading.
Sauer et al. [?] describe a loss reduction by leading edge modifications as seen in Figure 1.10. They introduced a bulb over the lowest 5% of the span. The rationale was to increase the strength of the suction leg of the

horseshoe vortex to prevent the impingement of the horse-shoe pressure side limb on the suction side of the blade. A 47% secondary loss reduction was achieved.

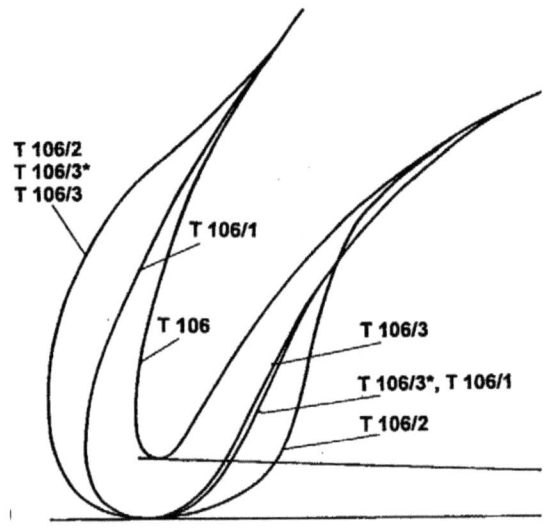

Figure 1.10: Leading Edge Modification [?]

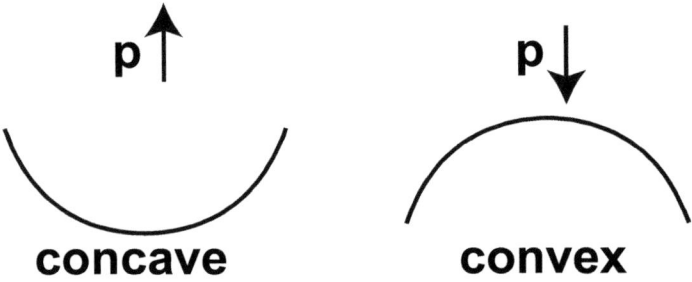

Figure 1.11: Effect of Streamline Curvature

1.1.7 Non-Axisymmetric End Wall Contouring

During the past decade emerging CFD capabilities have made it possible to design more complex 3-dimensional non-axisymmetric end walls. Non-axisymmetric profiling using such capabilities was first introduced by Rose [?]. His goal was to have a more homogenous pressure field at the vane rim, which should then required less turbine disk coolant massflow. The basic idea is to use streamwise curvature to control pressure. In order to balance the centripetal force created due to curvature a streamwise normal pressure gradient is developing. Concave curvature leads to an increase in local static pressure as the velocity is reduced and vice versa. Therefore, the actual magnitude of a perturbation is of lower significance than the actual curvature. See Figure 1.11 for illustration. This approach was later used to reduce the late cross passage pressure gradient in order to reduce secondary flow. Therefore, Harvey et al. [?] introduced an inverse approach to design non-axisymmetric end walls in order to reduce secondary loss. The end wall was parametrized by circumferential fourrier curves and splines in streamwise direction. Then a sensitivity matrix was constructed with secondary kinetic energy as the target function. The final end wall was then the result of a linear superposition of beneficial geometries. Hartland et al. [?] and Ingram et al. [?] investigated such designed non-axisymmetric end wall profiles in the Durham linear cascade and showed that secondary loss reductions of 24% could be attained. Brennan et al. [?] and Rose et al. [?] demonstrated an increase in stage efficiency of 0.4% from computations and 0.59%±0.25% from measurements for the Trent 500 HP model turbine using the same approach. Duden et al. [?] and Eymann et al. [?] investigated the combined effects of end wall contouring and blade thickening.

1.2 Research Objectives

The objective of this work is to increase the aerodynamic efficiency of a high work axial flow turbine using non-axisymmetric end wall profiling. For this purpose two non-axisymmetric end wall designs are compared to the axisymmetric end wall baseline case. It is thought likely that disk purge flow will affect the benefits of non-axisymmetric endwall profiling. More specifically; this thesis aims to hit the following targets:

- Investigate the steady and unsteady secondary loss mechanisms through extensive measurements in the first stage of a 1.5-stage axial flow turbine.

- Identify the improvements contributed by end wall profiling.

- Determine the sensitivity of the end walls to purge flow.

- Describe in detail the interaction of the rotor upstream purge flow and the rotor hub secondary flow structures, through detailed time-resolved CFD simulation.

- Come up with design recommendations for end wall profiling regarding purge flow.

1.3 Thesis Outline

Chapter 1 The literature review is given in section 1.1. The following topics are addressed: Secondary flows, tip leakage since the 1.5-stage configuration is unshrouded, unsteady flow interaction, rim seal purge flow, losses, secondary flow control and non-axisymmetric end wall profiling in particular.

Chapter 2 In this chapter the test rig and measurement technology are described. The new 'frozen flow' approach to calculate the three-dimensional vorticity vector from a single traverse plane is introduced. Finally the numerical methodology is explained at the end of this chapter.

Chapter 3 This chapter gives a detailed overview of the time-averaged and the time-resolved flow field analysis of the three end wall configurations without purge flow. Furthermore the relationship between secondary kinetic energy reduction and loss reduction is assessed. Finally the effect of contouring on the end wall pressure field is addressed.

Chapter 4 In this chapter the detailed analysis of the effects of suction and purge flow injection on the blade row secondary flow structures is presented for the axisymmetric baseline case. The purge flow mechanism itself is explained based on the time-resolved computational model.

1.3 Thesis Outline

Chapter 5 In this chapter the combined effects of purge flow injection and end wall profiling are shown. In the first part the three end walls are compared at an injection rate of 0.9% of the main flow. In the second part the sensitivity to purge flow for the second end wall design is presented based on two additional measurements at two different injection rates.

Chapter 6 In the last chapter the main conclusions and contributions are given as well as proposals for design future work.

2 Research Facility and Methods

2.1 Experimental Facility

2.1.1 Overview

The test facility used for the experiments reported is the axial research turbine facility LISA at the Laboratory of Energy Conversion of ETH Zurich. It is a continuously running, moderate-speed, low-temperature rig, which initially was designed to investigate unsteady effects in axial turbines.

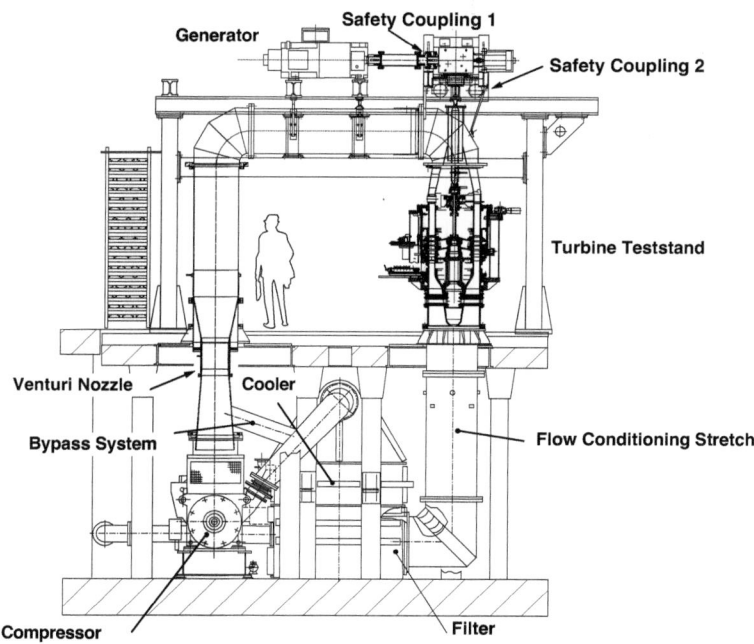

Figure 2.1: Schematic View of the LISA Research Turbine Test Rig

The facility extends over 3 floors. The air-loop of the facility is quasi-closed opening to atmosphere at exit of the turbine. A 1 MW radial compressor with a maximum pressure ratio of $\Pi_{c,max} = 1.5$ and a maximum massflow of 13 kg/s situated in the basement delivers the airflow. The operating point of the compressor and with it the turbine pressure ratio is controlled by adjustable inlet guide vanes. Before the flow enters the compressor, it goes through a calibrated venturi nozzle to measure the main massflow. At the exit of the compressor the flow passes through two water to air heat exchangers, which control the inlet total temperature to a constant value of $55°C$. Upstream of the turbine section is a 3m flow conditioning stretch to ensure a homogenous flowfield. Additionally, the flow undergoes an acceleration ahead of the turbine section in order to reduce the significance of remaining flow non-uniformities from upstream. On the top floor there is an angular gear box, which halves the rotational speed. A horizontal shaft connects the gearbox to a DC generator, which absorbs the turbine power and controls the rotational speed with an indicated accuracy of $\pm 0.02\%(\pm 0.5 rpm)$. A heat exchanger controls the inlet total temperature $T_{t,in}$ to an accuracy of $0.2K$. A torquemeter measures the torque on the rotor shaft.

2.1.2 1.5-Stage Turbine

Recently the existing 2-stage, shrouded turbine configuration, Sell et al. [?], was redesigned by Behr [?] as a 1.5-stage turbine with the following design objectives:

- Model of unshrouded, high-pressure gas turbine stage
- High stage loading
- Compressibility effects
- Low aspect ratio
- Realistic rotor exit flow field through a 2nd stator

The redesigned turbine consists of one highly loaded stage and a second vane. This results in maximized Mach numbers, stage loading and pressure ratio for the first stage. The presence of the downstream second stator row results, through upstream potential effects, in a representative stage exit

2.1 Experimental Facility

flow field. This adds in additional rotor stator interaction phenomena to be investigated.

Row		S1	R1	S2
Blade Count		36	54	36
Aspect Ratio (span/chord)	[−]	0.87	1.17	0.82
Solidity (chord/pitch)	[−]	1.27	1.41	1.34
Profile Stacking		LE	CoG	LE
Mach Number(averaged)	[−]	0.54	0.50	0.46
Reynolds Number	[−]	710k	380k	510k

Table 2.1: Characteristic Geometry and Performance Parameters of the 1.5-stage Turbine

With a compressor ratio limited to $\Pi_{c,max} = 1.5$, it is necessary to add a tandem de-swirl vane arrangement. This recovers the static pressure at the exit of the second stator back to the ambient level, in order to reach the intended turbine pressure ratio of $\Pi_{1.5} = 1.65$. The rotor nominal tip gap is 1% of the span with a variation between builds of less than 1% of the gap size. This ensures good repeatability. At exit of the first nozzle guide vane row the flow is compressible with an averaged exit Mach number of 0.53. In Table 2.1 the salient data of the 1.5-stage turbine is given.

Figure 2.2 shows the blade geometry at mid span as well as the relative positions of the traverse planes. The axial chord of the first vane is $50mm$. The axial gaps are 30% of the first vane axial chord. The first traverse (S1ex) is 16.6% of the first vane axial chord downstream of the first vane trailing edge and covers one stator pitch. It starts at −40% pitch relative to the first vane trailing edge. The second traverse plane (R1ex) is in the same circumferential position and covers one stator pitch as well. The axial position is 10.6% of the first vane axial chord downstream of the blade trailing edge.

2.1.3 Bypass and Flow Injection System

Figure 2.3 shows a schematic of the rotor upstream and downstream rim seals. Both rim seals are connected to the drum. The drum pressure level can be controlled by a valve which controls the ejection of massflow to the atmosphere. Before the air is released to ambient it passes a massflow

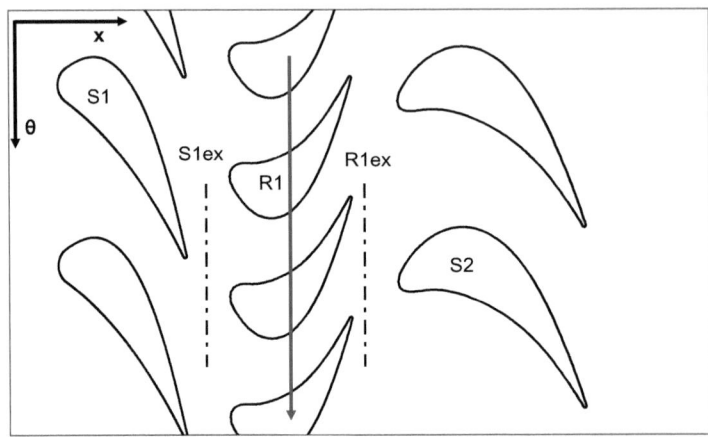

Figure 2.2: Illustration of Blade Geometry

measurement device. Previously the rig was operated in such a way that the pressure over the labyrinth seal in between the downstream rim seal and the drum was balanced. This should result in a negligible net massflow over the labyrinth seal. Under this circumstances the upstream rim seal is sucking a net massflow of 0.1%.

$$IR = \frac{\dot{m}_{by} - \dot{m}_{dr}}{\dot{m}_v} \cdot 100 \qquad (2.1)$$

For this investigation a new air-system was designed to provide the possibility of a positive net air flow through the rotor upstream rim seal. The air is bled off the primary air-loop upstream of the flow conditioning stretch. The bleed air passes through a venturi to measure the bypassed massflow. Finally the bypassed flow enters a plenum from where 10 plastic pipes lead the flow to 10 first nozzle guide vanes (S1). Through these vanes the flow labelled **B** enters the cavity in Figure 2.3. From the cavity underneath the vanes there are two leakage pathes indicated in Figure 2.3 as dotted arrows **P** and **S**. One path is through the upstream rim seal into the mainflow **P**. The rest of the gas which is ejected through the drum to the ambient after being measured in another venturi is called the secondary massflow **S**. With the assumption of zero net massflow over the labyrinth seal connecting to the downstream rim seal the injection or purge massflow can be calculated as the difference of the bypass massflow **B** and secondary massflow **S**.

2.2 Data Acquisition

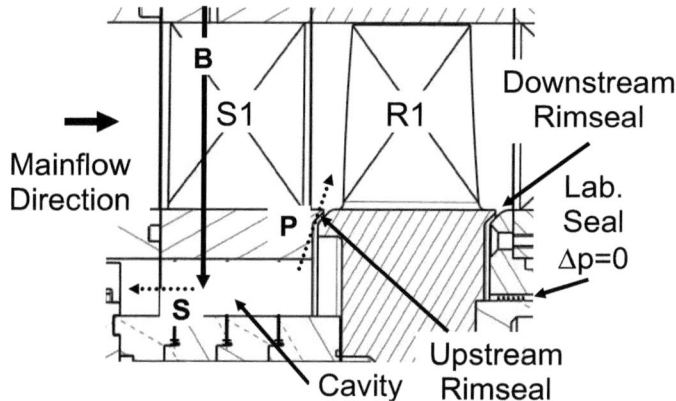

Figure 2.3: Illustration of Leakage Path

2.2 Data Acquisition

There are four independent main data acquisition chains, which include operating conditions, multi-channel pressure measurement system, vibration monitoring and probe measurement.

2.2.1 Operating Conditions

The measured operating conditions include pressures, torque, rotational speed, temperatures, humidity and massflow.

Pressure Two 16 channel difference pressure PSI modules are measuring the most important operating pressures. This includes the pressures used for the massflow calculations of the three massflow measurement devices, the inlet total and static pressure, the ambient pressure and the hub and tip pressure values at exit of each row. The inlet total pressure is measured with a pitot probe mounted at half the channel height on thin struts. These struts are circumferentially displaced relative to the traverse area. All other pressure values are measured with wall pressure tappings of $0.5mm$ diameter. The ranges of the two modules are $34.47kPa$ and $4.98kPa$ respectively. All pressures apart of the pressures of the massflow meter are measured against the same reference pressure to reduce uncertainties. As a reference pressure one of the first vane exit hub pressure tappings is chosen. Like that all sensors are operated in their optimum range.

Temperature and Humidity At the turbine inlet and exit four PT100 resistance thermometers are installed to determine the air temperatures. At the exit of the heat exchanger another PT100 thermometer measures the outlet temperature, which is the input to the inlet temperature control loop. Additional temperatures are measured at inlet of the massflow measurement devices as well as inside the cavity underneath the first vane. The latest is used to determine the temperature of the purge flow. In order to prevent condensation a humidity sensor is installed at the exit of the turbine. Another one is installed at the inlet of the main massflow meter to correct the density value for the actual humidity values.

Massflow Meters There are three massflow meters installed in the rig to measure the primary as well as the two secondary massflows. The primary massflow is measured with a calibrated Venturi nozzle. Since the upstream length does not conform with the minimum length defined in ISO 5167-3, the Venturi nozzle including the two upstream bends has been calibrated at Delft Hydraulics on a certified calibration rig for flow meters. Now the discharge coefficient of the nozzle can be determined as a function of the Reynolds number. The calculation of the massflow requires the absolute pressure, temperature and humidity of the flow at the exit of the nozzle as well as the pressure drop across the nozzle contraction.

The secondary massflow **S** described previously is measured with a ISA 1932 standard nozzle. The massflow calculation applied can be found in ISO 5167-3. The calculation of the massflow requires again the absolute pressure, temperature and humidity of the flow at the exit of the nozzle and the pressure drop across the nozzle contraction.

The bypass massflow **B** described further up is measured with a standard Venturi nozzle. The massflow calculation applied can be found in ISO 5167-3.

Torque The previous two stage configuration of the LISA rig had a twin shaft arrangement, which allows a independent assessment of the torque of each stage. The 1.5-stage turbine produces 20% more power than the previous two stage turbine. Therefore, the inner and outer shaft are coupled such that the torque is transmitted through each shaft according to its

2.2 Data Acquisition

stiffness. This results in a 3:1 torque split between the inner and outer shaft. The final torque reading is the sum of the two independent torque meters. The torque meters were specially designed and calibrated by Torquemeters Ltd. By measuring the phase shift of two cogwheels located at both ends of the calibrated torque shafts the angular deflection can be assessed and with it the torque. The shaft ratings of both torque meters are $1500Nm$ and $780Nm$ respectively.

2.2.2 Vibration

For safety reasons vibrations of the rotating parts are measured with displacement and acceleration sensors. The signals are processed and displayed by a Schenk vibration diagnostic module (VibroControl 4000). It indicates the displacement amplitudes and vibration velocities. The data is additionally displayed and logged on an independent personal computer. Three acceleration sensors are dedicated to monitor the vibrations of the gear box. One is used to measure the generator vibrations. The remaining four are located at the turbine bearings. If the set limits are exceeded an emergency shut down is initiated immediately.

2.2.3 Multi-Channel Pressure Measurement System

The multi-channel pressure measurement system consists of two Digitial Sensor Array Enclosures (DSAENCL 3200) with a Windows Embedded operating system which allows an Ethernet connection. The first enclosure is equipped with 3 ZOC16TC modules each containing 16 $10"H_2 0$ range pressure transducers. Additionally it contains two Servo Pressure Calibrators SPC 3000 and the Control Pressure Module CPM 3000 for the sensor calibration. The second enclosure contains 7 ZOC16TC modules each containing 16 $5PSI$ range pressure transducers. The data acquisition as well as the conversion to engineering units is done through the Scani-Valve DSMLINK V2.98 software. At the beginning of each campaign a full range calibration was performed and the new calibration coefficients were stored to the enclosures. Additionally every day before measuring a zero calibration is executed to remove a possible offset. At the same time the new zero offset is compared to the previous value to detect a possible drift.

2.3 Vane Instrumentation

In order to investigate the effects of non-axisymmetric end wall profiling a set of instrumentation is installed in the first vane row. It includes surface pressure measurements, end wall pressure measurements as well as a number of probe accesses into one vane passage. This instrumentation is only included in the two non-axisymmetric cases.

2.3.1 Vane Surface Pressure Measurements

The instrumentation of turbine vanes with pneumatic pressure tappings is still an important measurement to validate CFD results in turbomachinery flows. With this instrumentation the aerodynamic loading of the vane can be determined. Data by Rose et al. [?] has shown that end wall profiling changes the loading close to the end walls. Therefore, it has been decided to put two planes of pneumatic tappings close to the hub at **3%** and **5%** span and close to the tip end wall at **95%** and **97%** span. A fifth plane was added at **50%** span. Each plane consists of 30 tappings. The tappings have been machined into 5 different individual vanes. The tapping size is $0.5mm$ in diameter.

2.3.2 Vane End Wall Pressure Measurements

Non-axisymmetric end wall profiling strongly influences the static pressure field on the end walls. Therefore, a segment consisting of 3 vanes has been instrumented with **96** (8 tappings in circumference and 12 along the axial chord) pneumatic tappings $0.5mm$ in diameter on each end wall which were CNC machined. The distribution of the measurement points can be seen in Figure 2.4

2.3.3 Vane Intra-Row Traversing

In order to study the evolution of the secondary flows within the first vane a number of probe accesses have been designed. Each probe access hole is $4mm$ in diameter.
This limits the number of possible accesses in the circumferential direction. The possible number would have resulted in a to coarse mesh. In order to allow greater resolution an approach using replaceable inserts was adopted.

2.3 Vane Instrumentation

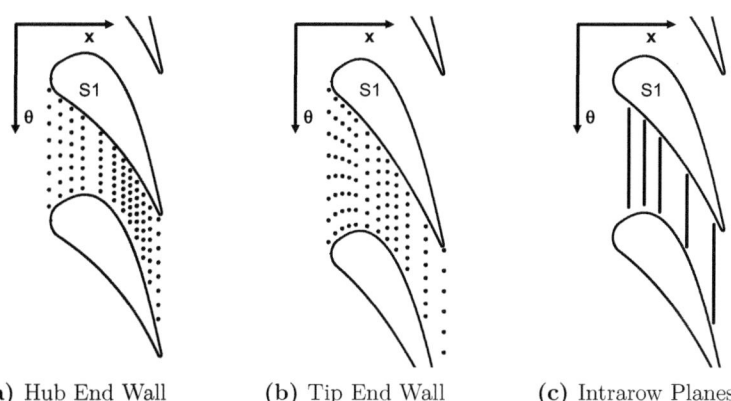

(a) Hub End Wall (b) Tip End Wall (c) Intrarow Planes

Figure 2.4: Pneumatic Tapping Distribution on Hub and Tip End Wall and Intra-Row Traverse Planes

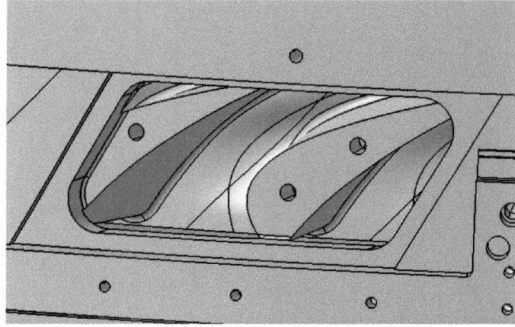

Figure 2.5: Pocket in Vane One Segment to Install Inserts

A pocket is machined into a 3 vane segment as seen in Figure 2.5. Afterwards 3 brass inserts containing the probe accesses are mounted into the pocket tripling the resolution in circumferential direction. The left side of Figure 2.6 shows the insert with the end wall profiling. In order to seal the probe accesses a plug containing an o-ring is inserted as seen on the right side of Figure 2.6. The plugs are machined together with the insert. This guaranties a smooth end wall surface. As the end wall is non-axisymmetric it is crucial to maintain the angular position of the plug. Therefore, the head of the plug has a square shape, which is constrained by a groove machined into the insert. The inserts contain 5 traverse planes inside the passage. The positions in percentage axial chord are given in Table 2.2 and shown in Figure 2.4. After each radial traverse the probe has to be manually repositioned. However, with the current design the process is very effective.

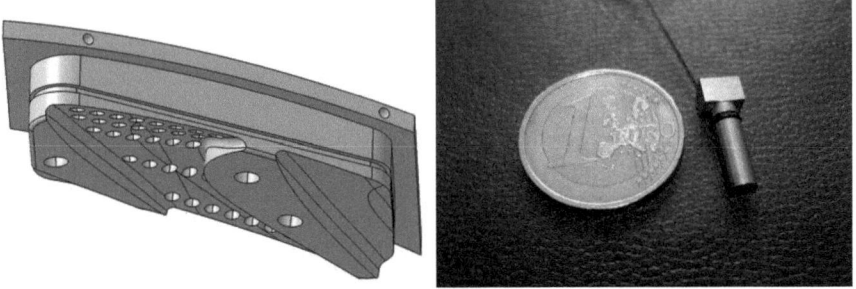

Figure 2.6: Insert (left) and Sealing Plug (right)

| 14% | 28% | 42% | 66% | 90% |

Table 2.2: Intra-row Axial Stations

2.4 Probe Technology

2.4.1 Introduction

The unsteady flow field in a turbomachine can be measured by a number of different technologies. First by non-intrusives such as laser doppler anemometry (LDA) or particle image velocimetry (PIV). Another group are the intrusive techniques such as hot-wire anemometry and fast response pneumatic probes. Non-intrusive techniques have the advantage of not disturbing the flow field. However, the fast response probe has the advantage of the possibility to measure the time-resolved local total and static pressure field in addition to the 3D velocity field. The pressure information is essential within this measurement campaign in order to quantify the loss reductions due to the non-axisymmetric end wall profiling.

2.4 Probe Technology

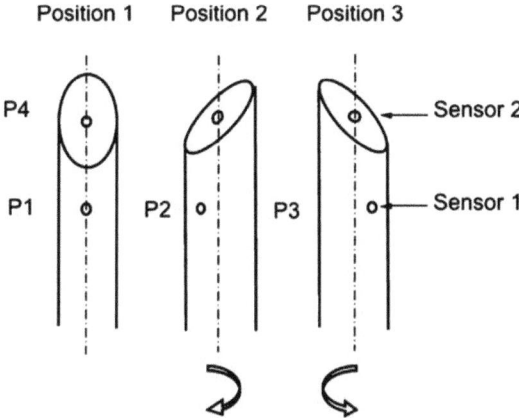

Figure 2.7: Measurement Concept of a Virtual 4-Sensor Probe

Fast-Response Aerodynamic Probe [FRAP] The fast response aerodynamic probe technology has been developed over the last two decades at the Laboratory of Energy Conversion at ETH Zurich. For details on the fundamentals and first measurements refer to Gossweiler [?] or for details on further measurements to Kupferschmied [?]. The blockage effects of the probe are minimized as a result of the miniaturization of the probe heads down to $1.8mm$ in diameter as seen in Figure 2.8. The probe head is cylindrical with an inclined, curved tip. The probe tip has been developed and applied to two complementary single-sensor FRAP's by Pfau [?]. Schlienger [?] further developed the design to a 2-sensor FRAP. The probe is capable of capturing unsteady flow features up to frequencies of $48kHz$ based on measurements including total and static pressures, flow yaw and pitch angles and Mach number. The frequency bandwidth of the temperature is limited to a frequency of $10Hz$. However, the influence of the measured temperature on the velocity is negligible. The 2-sensor FRAP is operated in a virtual-4-sensor mode to measure 3-dimensional, time-resolved flow properties. The concept is graphically displayed in Figure 2.7. The yaw sensitivity is gained from sensor 1 by turning the probe $-42°$ and $42°$ according to is angular position. Therefore, each radial measurement point consists of 3 different angular positions. As a result of the virtual approach only the periodic part of the time-resolved yaw angle information can be resolved due to phase locking. The pitch sensitivity is the result of the second sensor located on the curved surface.

Pneumatic Probes Additionally to the fast-response aerodynamic probe (FRAP) a pneumatic 5-hole cobra shaped probe is used with a tip diameter of $0.9mm$ as shown in Figure 2.8. The main advantage of the cobra shaped probe is a large distance of the probe shaft from the measurement volume.

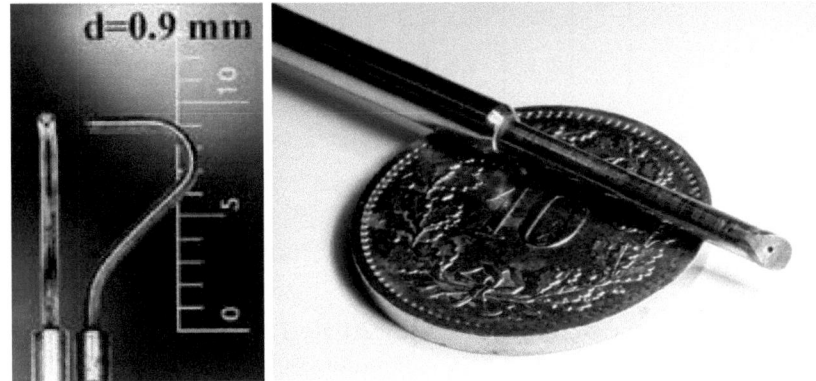

Figure 2.8: Cobra Shaped 5-Hole Probe (left) and FRAP (right)

Through the simplicity of the measurement chain the aerodynamic probes deliver good reference data for the FRAP probe. For the measurements within the first vane row a pneumatic 4-hole probe with an elliptical head is used. The extra long $3mm$ shaft allows access over the whole span and reduces blockage effects within the passage.

2.4.2 Calibration

There are two facilities dedicated to the probe calibration.

Freejet Facility Before each campaign all probes are calibrated in the freejet facility for yaw and pitch angle and total and static pressure at a given Mach number. The axisymmetric freejet has a uniform velocity profile and a turbulence level of 0.3%. A detailed description of the facility can be found in Kupferschmied [?]. In the freejet facility the probes can be calibrated for a yaw angle range of $\pm 180°$ and a pitch angle range of $\pm 36°$. The maximum Mach number is around 0.9.

2.4 Probe Technology

Sensor Calibration Facility The pressure sensors of the fast-response aerodynamic probe have to be calibrated for pressure and temperature. For the calibration an oven is used which provides especially accurate and stable temperatures. At the same time a back pressure is applied within the probe cavity, provided by the same very accurate pressure source as applied during measurements.

Pneumatic 5-Hole Probe The 5-hole probe has been calibrated for a yaw angle range of ±14° and a pitch angle range of ±20° with steps of 2°.

Inlet	S1ex	R1ex	S2ex
0.14	0.53	0.25	0.46

Table 2.3: Calibration Mach Numbers 5-Hole Probe and FRAP

The 5-hole probe was calibrated at four different Mach numbers given in Table 2.3 corresponding to the average inlet Mach number and the Mach numbers at exit of each row. Out of a calibration a set of coefficients is determined. The Equations to evaluate the sensitivity coefficients K_i are given in Equation 2.2:

$$K_\varphi = \frac{p_2 - p_3}{q}, K_\gamma = \frac{p_4 - p_5}{q}, K_t = \frac{p_t - p_1}{q}, K_s = \frac{p_t - p_s}{q} \qquad (2.2)$$

with

$$q = p_1 - \frac{1}{4}\sum_{i=2}^{5} p_i \qquad (2.3)$$

Pneumatic 4-Hole Probe The 4-hole probe was used to measure inside the first vane row because it requires a smaller measurement access than the 5-hole probe. The probe was calibrated for a yaw angle range of ±14° and a pitch angle range of ±20° with steps of 2° and for five different Mach numbers as given in Table 2.4. As the Mach number rises through the first row from $M = 0.14$ to $M = 0.53$ the probe was calibrated over the whole range from 0.1 to 0.6.
The calculation of the sensitivity slightly changes as seen in Equation 2.4.

$$K_\varphi = \frac{p_2 - p_3}{q}, K_\gamma = \frac{p_1 - p_4}{q}, K_t = \frac{p_t - p_1}{q}, K_s = \frac{p_t - p_s}{q} \qquad (2.4)$$

| 0.1 | 0.3 | 0.4 | 0.5 | 0.6 |

Table 2.4: Calibration Mach Numbers Pneumatic 4-Hole Probe

with

$$q = p_1 - \frac{1}{2}\sum_{i=2}^{3} p_i \qquad (2.5)$$

FRAP The sensor calibration was done over a temperature range from $30°C$ to $65°C$ in steps of $5°C$. The pressure sensitivity was determined by changing the back pressure in between $1kPa$ and $57kPa$ in steps of $8kPa$. The pressure sensitivity of both sensors is $0.95mV/kPa$, while the temperature sensitivity is $2.6mV/°C$.
The FRAP has been calibrated for a yaw angle range of $\pm 80°$ and a pitch angle range of $\pm 20°$ with steps of $5°$. The sensitivity coefficients for FRAP are the same as for the pneumatic 4-hole probe as given in Equation 2.4.

2.4.3 Traversing System and Resolution

Traversing System The test rig is equipped with a 3-axis fully automated traversing system. The radial movement and the turning around the probe axis is enabled by two stepper motors. The turbine is equipped with 3 movable casing rings, which slide on radial seals. A third motor connected to a Heidenhain encoder enables the circumferential movement of the probe. This design explained in Schlienger [?] minimizes the flow disturbances allowing continuous area traversing in one plane with a probe access of only $10mm$ in diameter. The accuracies and ranges of the traversing system components are given in Table 2.5.

Traversing Axis	Range	Accuracy
Circumferential	30°	0.002°
Radial	150mm	0.1mm
Yaw Angle	360°	0.003°

Table 2.5: Range and Accuracies of the Traversing System

2.5 Data Reduction

Resolution The spatial resolution of the measurement grid for a pneumatic area traverse consists of 40 radial and 40 circumferential points (covering one stator pitch) with a radial clustering near the end walls. The first traverse is repeated in order to check for the periodicity. With the FRAP probe the radial point closest to the hub end wall couldn't be reached for safety reasons, resulting in only 39 radial points. The relative frame resolution can be evaluated as follows. The data logging is done for $2s$ with a sampling rate of $200kHz$. At $2700rpm$ this is equivalent to 90 rotor revolutions. With 54 rotor blades this results in 4869 blade passing events. This leads to about 82 samples per rotor pitch. The Nyquist criterion leads to a halving of the resolution.

2.5 Data Reduction

2.5.1 Pneumatic Probe Data

The pneumatic post processing can be divided into two steps:

1. Data-Processing
2. Visualization

Data-Processing The operational data and probe data are logged with the same time stamp. Then using the calibration coefficients and the 5-hole pressures the flow angles, total and static pressure are calculated with bivariable polynomials given in the following Equations:

$$\varphi = \sum_{i=0}^{6} \sum_{j=0}^{6} k_{\varphi,ij} K_\varphi{}^i K_\gamma{}^j \qquad (2.6)$$

$$\gamma = \sum_{i=0}^{6} \sum_{j=0}^{6} k_{\gamma,ij} K_\varphi{}^i K_\gamma{}^j \qquad (2.7)$$

$$K_t = \sum_{i=0}^{6} \sum_{j=0}^{6} k_{t,ij} \varphi^i \gamma^j \qquad (2.8)$$

$$K_s = \sum_{i=0}^{6} \sum_{j=0}^{6} k_{s,ij} \varphi^i \gamma^j \qquad (2.9)$$

k are the polynomial coefficients determined during the calibration. The angles are still in the probe relative system. Therefore, the traversing set angles have to be added. The angle convention is as follows: The yaw angle is positive in the rotational direction of the turbine, the pitch angle is positive for fluid moving towards the tip. The yaw angle is given relative to the axial direction. Then a number of parameters is calculated as given in Table 2.6 with their definition. For the speed of sound the average temperature from the FRAP measurement at the traversing plane is applied. The error on the speed of sound comparing it to the value based on the temperature measured with the FRAP probe is in the range of $\pm 1\%$ and is therefore not a source of significant error. As the turbine is open to atmosphere at the exit, the pressure data has to be corrected for daily variations. Therefore, throughout this work only pressure ratios PR are used. For the first vane row the total pressure loss coefficient is evaluated. During the measurements an additional traverse is taken which is exactly one pitch apart of the first traverse. This traverse is then used to calculate the cells around the edge.

Mach Number	$M = \sqrt{\frac{2}{\gamma-1}\left[\left(\frac{p_t}{p}\right)^{\frac{\gamma-1}{\gamma}} - 1\right]}$
Absolute Velocity	$C = M \cdot \sqrt{\gamma RT}$
Axail Velocity	$C_x = C \cdot \cos\varphi \cdot \cos\gamma$
Radial Velocity	$C_r = C_x \cdot \tan\gamma$
Absolute Velocity	$C_\varphi = C_x \cdot \tan\varphi$
Total Pressure Ratio	$PR\_t = \frac{p_t}{p_{t,in}}$
Static Pressure Ratio	$PR\_s = \frac{p}{p_{t,in}}$
Total Pressure Loss Coefficient (Vane 1)	$Y = \frac{p_{t,in} - p_t}{p_{t,in} - p}$

Table 2.6: Parameters Calculated During Data-Processing

The last step in the data-processing algorithm is to do the area and massflow averaging. Generally non-convective flow quantities are expressed as area-averages, while convected quantities are massflow averaged. As the measurement grid is discrete, the integration is a numerical one. Therefore, the flow quantities are cell centered and multiplied by the cell area or massflow depending on the averaging process. Depending on the direction of the integration the result is a circumferential or radial average. The average of the whole pitch is achieved by a two directional integration.

2.5 Data Reduction

Area Averaging:

$$dA = r \cdot dr \cdot d\theta \qquad (2.10)$$

$$\overline{f} = \frac{\Sigma f \cdot dA}{\Sigma dA} \qquad (2.11)$$

Mass Averaging:

$$d\dot{m} = \rho \cdot C_x \cdot dA \qquad (2.12)$$

$$\overline{f} = \frac{\Sigma f \cdot d\dot{m}}{\Sigma d\dot{m}} \qquad (2.13)$$

Visualization In the second step the preprocessed data can be plotted as line plots or as contour plots of the whole area. Furthermore the plotting range can be chosen.

2.5.2 FRAP 2-sensor

The time-resolved data is reduced with the in house software package HERKULES. The program is extensively explained in Schlienger [?]. The data processing is divided into four steps:

- **Step 1:** Generation of GEO and RIG File
- **Step 2:** Raw Data Preprocessing
- **Step 3:** Offset Gain Calibration, Evaluation of Flow Quantities, Phase-Lock Averaging
- **Step 4:** Post-Processing and Visualization

Step 1: The GEO file contains the point coordinates as well as the yaw setting angle values of each measurement point. The RIG file contains the inlet total pressure and exit static pressure at the beginning an the end of each traverse to calculate the non-dimensional pressure quantities.

Step 2: In this step the raw signal is read and phase-locked to the once per revolution trigger signal. In this step the data is reduced to only the first three blade passages. With a resolution of 82 samples per blade passage this results in 246 data points. Times 85 revolutions, this results in 20910 samples per measurement point.

Step 3: With the sensor model and the sensor calibration coefficients the voltage signals are converted into pressure and temperature quantities. 2.14 and 2.15 are the Equations to calculate the pressure difference and temperature of the two sensors.

$$\Delta p = \sum_{i=0}^{m} \sum_{j=0}^{n} a_{p,ij} U^i U_e^j \qquad (2.14)$$

$$T = \sum_{i=0}^{m} \sum_{j=0}^{n} a_{T,ij} U^i U_e^j \qquad (2.15)$$

In order to remove the sensor drift effects investigated by Kupferschmied [?] an Offset-Gain calibration is executed before and after a radial traverse. A detailed description can be found in Schlienger [?]. The new offset and gain values are applied to the calculated pressure difference and temperature. Afterwards the flow angles, total and static pressures are calculated with bivariable polynomials as given in Equations (2.6-2.9). Then an out-of-calibration filter is applied, which was introduced by Behr [?]. The next step is the phase-lock averaging. Finally all relevant flow quantities are evaluated including turbulence values (Porecca et al. [?]) and then non-dimensonalized if necessary.

Step 4: The last step includes the calculation of additional quantities such as streamwise vorticity or secondary kinetic energy. The process is described in subsections 2.5.3 and 2.5.4. Additionally there is variety of visualization possibilities. Behr [?] has included the option to visualize unequal stator-rotor blade counts. For more detail on the visualization refer to Schlienger [?] or Behr [?].

2.5.3 Secondary Kinetic Energy

As secondary kinetic energy is the prime design parameter for the non-axisymmetric end walls tested in this work, the option to calculate it from

2.5 Data Reduction

the pneumatic and time-resolved data is indispensable. The secondary kinetic energy is the square value of the secondary velocity vector. The secondary velocity is calculated as the difference vector between the measured velocity vector and a defined primary velocity vector. In this work a mass flow averaging process in radial and circumferential direction has been used as reported by Germain et al. [?]. First a radial window averaging is done. The chosen window size is a quarter span height. Towards the end walls the window is shrinking in relation to the proximity of the measurement point as seen in Figure 2.9

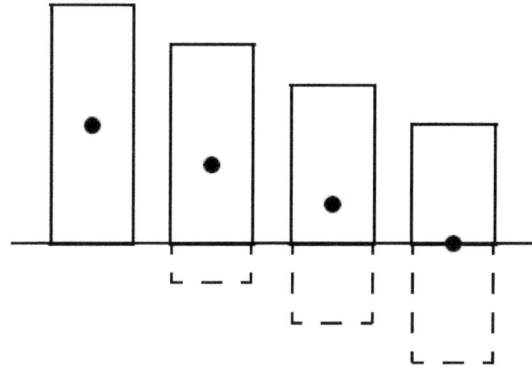

Figure 2.9: Window Averaging Concept for Secondary Kinetic Energy Calculation

The radial window averaging is only applied to the axial and circumferential velocity components $(i = \varphi, x)$.

$$\overline{C_i}(\theta) = \langle C_i(r,\theta) \rangle_r \qquad (2.16)$$

By subtracting the averaged quantities from the velocity components a first secondary quantity is derived.

$$C'_i(r,\theta) = C_i(r,\theta) - \overline{C_i}(\theta) \qquad (2.17)$$

These are then circumferentially mass-averaged as well as the radial velocity component.

$$\overline{\overline{C_i}}(r) = \langle C'_i(r,\theta) \rangle_\theta \qquad (2.18)$$

$$\overline{C_r}(r) = \langle C_r(r,\theta) \rangle_\theta \tag{2.19}$$

Then the final secondary velocity components are calculated.

$$C_i''(r,\theta) = C_i'(r,\theta) - \overline{\overline{C_i}}(r) \tag{2.20}$$

$$C_r'(r,\theta) = C_r(r,\theta) - \overline{C_r}(r) \tag{2.21}$$

Finally the secondary kinetic energy is calculated.

$$SKE = C_x''^2 + C_\varphi''^2 + C_r'^2 \tag{2.22}$$

2.5.4 Streamwise Vorticity

The streamwise vorticity is related to secondary flows as it introduces flow perpendicular to the primary flow direction, defined by the circumferentially mass-averaged velocity profile. Therefore, a reduction in streamwise vorticity normally leads to a reduction in secondary flow. As the goal of non-axisymmetric end wall profiling is to reduce the secondary flow, the streamwise vorticity becomes an important flow quantity to evaluate.

The streamwise vorticity ω_s is finally calculated as the scalar product of the vorticity $\underline{\omega}$, whose components are given in Equations (2.24-2.27) and the primary flow vector $\underline{C}_s$.

$$\omega_s = \underline{\omega} \cdot \underline{C}_s \tag{2.23}$$

In this work the primary flow direction is derived from circumferential mass-averaging of all three velocity components and is therefore a radial varying quantity.

$$\omega_x = \frac{1}{r}\left(\frac{\partial}{\partial r}(rC_\theta) - \frac{\partial C_r}{\partial \theta}\right) \tag{2.24}$$

$$\omega_r = \frac{1}{r}\frac{\partial C_x}{\partial \theta} - \frac{\partial C_\theta}{\partial x} \tag{2.25}$$

$$\omega_\theta = \frac{\partial C_r}{\partial x} - \frac{\partial C_x}{\partial r} \tag{2.26}$$

$$\tag{2.27}$$

2.5 Data Reduction

Frozen Flow Approach To calculate the three-dimensional vorticity vector the axial gradients of all three velocity components are needed. By measuring one plane as it is standard in the LISA rig this information is not available. With the frozen flow approach the axial gradients are estimated using the time-resolved data with the assumption that the flow structures are frozen within one timestep. Using this assumption the following approximation of the axial derivative can be used. As for example for the radial velocity;

$$\frac{\partial C_r}{\partial x} \approx \frac{\partial C_r}{C_x \cdot \partial t} \quad (2.28)$$

However, this approach is only correct if the flow direction varies within a very small angle of $\pm 10°$ with respect to the axial direction. For larger angle variation it is necessary to interpolate in the circumferential direction using the circumferential displacement $d\theta$ of a fluid particle over one timestep t. So if the fluid particle is travelling in positive circumferential direction it is displaced by $-d\theta$ relative to the traverse Tr in the preceding timestep. Therefore, an interpolation within the measurement grid consisting of the radial traverses Tr is required to determine the velocity at the same circumferential position. Equation 2.28 is then rewritten as Equation 2.29. An illustration is given in Figure 2.10.

$$\frac{\partial C_r}{\partial x} \approx \frac{C_r(t,\theta) - C_r(t+1, \theta + d\theta)}{C_x \cdot dt} \quad (2.29)$$

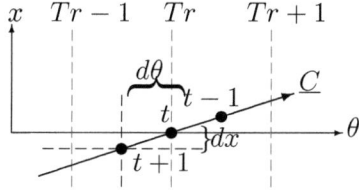

Figure 2.10: Illustration for Circumferential Interpolation

Validation In this paragraph the newly developed approach is validated against two other procedures.

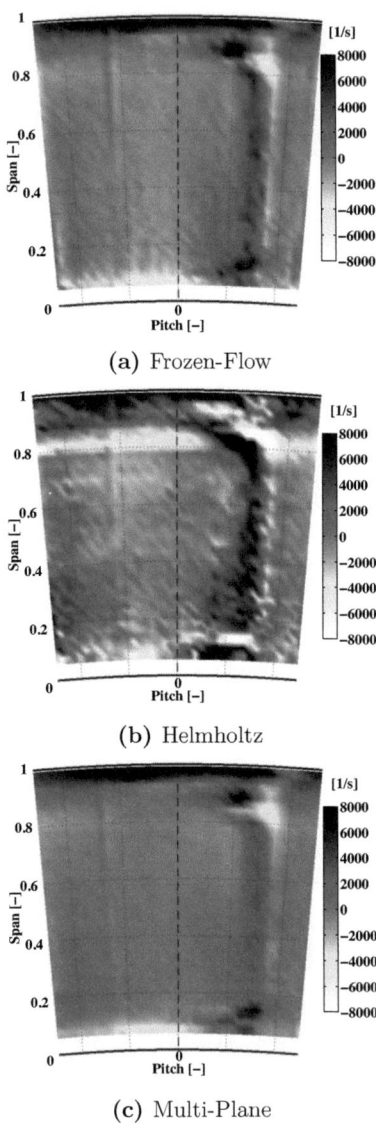

Figure 2.11: Validation of Different Approaches to Calculate the Streamwise Vorticity at S1ex

One approach to determine the axial velocity derivates needed to calculate the three dimensional vorticity vector is to measure two axial planes which are displaced by a short axial distance. However, this means a doubling of the measurement time as two traverse planes have to be measured. For the purpose of validation two planes $0.5mm$ apart have been measured with FRAP at the exit of the first vane row. Figure 2.11 (c) shows the streamwise vorticity evaluated with this approach.

Gregory-Smith et al. [?] have introduced the concept of using the incompressible Helmholtz Equation 2.30 in a linear low speed cascade.

$$\underline{C} \times \underline{\omega} = \frac{1}{\rho} \nabla p_t \qquad (2.30)$$

With this approach the measurement of two planes can be avoided by using the total pressure gradients within one traverse plane. However, this is an incompressible approach. It does not work at the exit of the first vane of this turbine if Figure 2.11 (b) and (c) are compared. With the Helmholtz approach there is a band of negative vorticity at 80% span which seems to be unphysical. Furthermore, there is an unphysical result in the hub secondary flow region.

If Figure 2.11 (a) and (c) are compared one can see an acceptable level of agreement between the Frozen Flow approach and the direct measurement of the axial derivatives. So it can be concluded that the new frozen flow approach gives a much better result than the previous Helmholtz approach if the flow is incompressible. Furthermore the approach gives a comparable result to the multiplane measurement with half the experimental time required.

2.6 Uncertainty Analysis

The uncertainty analysis presented in this document is done using the "Guide of Uncertainty in Measurements" (GUM) [?]. It is a standardized method, which first converts all uncertainty information in probability distributions. In case of correlated parameters, cross-correlation coefficients are needed to evaluate their combined uncertainty contribution. The methods for the evaluation of the results and the associated uncertainty are defined in detail and should be followed to achieve transparency and reliability. In DIN 1319-3 section 4.2 it is recommended to divide the evaluation process into four steps:

1. Development of the model, which describes the measurement problem in the form of mathematical equations.
2. Preparation of the input data and of additional information.
3. Calculation of the results and the associated standard uncertainties with the given input quantities and the given model.
4. Notification of the complete measurement result including the measurement uncertainty.

In step 2 the knowledge is converted into a probability distribution. From the distribution the expectation value as well as the variance can be calculated. The GUM method differentiates between two kinds of input data.

- Type A: Observed statistical data collected during a measurement
- Type B: Non-statistical data, which are known prior to the measurement

The best practice is to do it as simply as possible. If only the two limits are known, it is suggested to use a rectangular probability distribution. If the input is the result of a measurement the normal distribution is recommended. The standard uncertainty u of the rectangular distribution with a known half-width a is calculated with

$$u(x_i) = \frac{a}{\sqrt{3}} \qquad (2.31)$$

In step 3 the partial derivatives of the model function y are calculated to obtain the sensitivity c_i of every input quantity x_i

$$c_i = \frac{\partial y}{\partial x_i} \qquad (2.32)$$

The uncertainty contribution $u_i(y)$ is calculated by a multiplication of the sensitivity coefficient c_i with the standard uncertainty of the input quantity $u(x_i)$.

$$u_i(y) = c_i \cdot u(x_i) \qquad (2.33)$$

The law of the error propagation is given as follows:

2.6 Uncertainty Analysis

$$u^2(y) = \sum_{i=1}^{n} u_i^2(y) \left[+2 \cdot \sum_{i=1}^{n-1} \sum_{j=i+1}^{n} r(x_i, x_j) \cdot u_i(y) \cdot u_j(y) \right] \quad (2.34)$$

The term in the square bracket in Equation 2.34 is only necessary if two input quantities x_i and x_j are correlated. The GUM method uses a special term called the *expanded uncertainty*. It was introduced to have a kind of interval to be able to compare to other specifications or other kind of limits. The *expanded uncertainty* U consists of the calculated standard uncertainty $u(y)$ multiplied by a coverage factor k.

$$U = k \cdot u(y) \quad (2.35)$$

Normally a coverage factor of $k = 2$ is used which corresponds to a confidence level of 95%. Behr [?] has calculated the error propagation for the pneumatic 5-hole probe technique using the calibration polynomial model. Table 2.7 gives the expanded uncertainty for the probe at two representative Mach numbers with an expansion factor of $k = 2$. The pressure values are normalized with a standard dynamic head, which is based on the standard atmosphere.

flow quantity	$M = 0.25$	$M = 0.5$
yaw angle	0.3°	0.3°
pitch angle	0.3°	0.3°
total pressure	1.8%	0.6%
static pressure	2%	1%

Table 2.7: Expanded Uncertainty of the Pneumatic 5-Hole Probe

flow quantity	*Accuracy*
yaw angle	0.5°
pitch angle	0.7°
total pressure	1.0%
static pressure	1.2%

Table 2.8: Averaged Measurement Accuracy of FRAP System with 2-Sensor FRAP

Table 2.8 gives a general averaged accuracy of the FRAP system. In order to derive the *expanded uncertainty* of the mass-flow meters and the turbine efficiency the two model functions are needed. Equation 2.36 shows the through flow Equation to calculate the massflow $\dot{m}_i$ for each massflow meter

$$\dot{m}_i = \frac{\pi}{4}d^2\frac{C_d}{\sqrt{1-\beta^4}}\epsilon\sqrt{2\Delta p\rho} \qquad (2.36)$$

Table 2.9 gives the full-scale uncertainties of all measurement devices within the measurement chain. All this input data is converted into rectangular distributions with a half-width corresponding to the indicated uncertainty.

Measurement Device	Parameter	Range	Uncertainty
ScaniValve ZOC16	Δp	$0\ldots35kPa$	$\pm0.065\%FS$
PSI 9016 diff. press.	Δp_v	$0\ldots5kPa$	$\pm0.15\%FS$
PSI 9016 diff. press.	Δp	$0\ldots35kPa$	$\pm0.065\%FS$
Keller X33 abs. press.	p_{atm}	$0.8\ldots1.2bar$	$\pm0.05\%FS$
resistance thermometer	T	$0\ldots60°C$	$\pm0.3\%FS$
relative humidity sensor	H	$0\ldots100\%$	$\pm1.0\%FS$

Table 2.9: Accuracy of Measurement Devices

Input Quantity	Parameter	Relative Uncertainty $\Delta x/x$
discharge coefficient	C_d	$\pm0.16\%$
throat diameter	d	$\pm0.002\%$
inlet diameter	D	$\pm0.001\%$

Table 2.10: Accuracies Related to the Main Venturi

For the main venturi the uncertainty analysis reveals an relative *expanded uncertainty* (k=2) of $\pm0.22\%$ at an average massflow of $11.8kg/s$. The uncertainty value of the discharge coefficient C_d contributes 53% to the massflow uncertainty, while the pressure difference uncertainty is responsible for another 45%. The remaining 2% are contributed by the other input parameters.

The efficiency Equation is given in Equation 2.37 where IR is the injection rate as defined in Equation 2.1.

2.6 Uncertainty Analysis

$$\eta_{tt} = \frac{\frac{\omega \cdot M}{\dot{m}_v \cdot c_p \cdot T_{t,in}}}{1 - \left(1 - \frac{IR}{100}\right) \cdot \left(\frac{p_{t,R1ex}}{p_{t,in}}\right)^{\frac{\gamma-1}{\gamma}} - \frac{IR}{100} \cdot \left(\frac{p_{t,R1ex}}{p_{t,cav}}\right)^{\frac{\gamma-1}{\gamma}}} \quad (2.37)$$

Input Quantity		Standard Uncertainty $u(x_i)$	Uncertainty Contribution
rotational speed	ω	$\pm 0.06\ [rad/s]$	1.1%
torque	M	$\pm 0.866\ [Nm]$	17.0%
main massflow	$\dot{m}_v$	$\pm 0.013\ [kg/s]$	29.6%
secondary massflow	$\dot{m}_{dr}$	$\pm 0.0003\ [kg/s]$	0.0%
bypass massflow	$\dot{m}_{by}$	$\pm 0.0009\ [kg/s]$	0.0%
inlet total temperature	$T_{t,in}$	$\pm 0.1\ [K]$	2.4%
inlet total pressure	$p_{t,in}$	$\pm 18.4\ [Pa]$	4.0%
exit total pressure	$p_{t,R1ex}$	$\pm 45\ [Pa]$	45.8%
cavity pressure	p_{cav}	$\pm 25\ [Pa]$	0.0%

Table 2.11: Accuracies Related to the Total-to-Total Efficiency

In order to calculate the *expanded uncertainty* of the total-to-total efficiency in the case of purge flow injection further uncertainty input parameters are needed. The related standard uncertainties are given in Table 2.11 as standard uncertainties (k=1). The exit total pressure value $p_{t,R1ex}$ is calculated as a mass-averaged quantity from a 5-hole probe measurement at exit of the rotor. The quantification of the uncertainty contributed by not measuring the lowest 4% of the span corresponding to 3% of the total massflow revealed a $20Pa$ standard uncertainty. With the standard uncertainty of the 5-hole probe total pressure reading of $40Pa$ this results in a combined standard uncertainty of $45Pa$ for the exit total pressure $p_{t,R1ex}$. The calculated absolute *expanded uncertainty* (k=2) of the total-to-total efficiency is $\pm \mathbf{0.37\%}$. The uncertainty of the exit total pressure shows with 45.8% the largest contribution to the overall uncertainty, while the main massflow measurement and torque measurement are responsible for 29.6% and 17.0% respectively. The remaining 7.6% are contributed by the other input quantities. The contribution of the secondary massflows and the cavity pressure is negligible.

In order to quantify the expanded uncertainty of the delta total-to-total efficiency of two measurements the following assumptions are made. In this work all efficiency measurements were executed with the same probe and the same calibration files. Therefore, the uncertainty of the calibration is not considered. For relative measurements the contribution of the discharge coefficient on the massflow uncertainty can be neglected as well. The resulting expanded uncertainty (k=2) for a change in total-to-total efficiency is ±**0.32%**. However, this only holds for a back to back measurement using the same probe and calibration coefficients.

2.7 CFD

2.7.1 Steady CFD

The first vane of the axisymmetric baseline turbine hasn't been equipped with the instrumentation described in section 2.3. In order to be able to compare the measured results of the non-axisymmetric cases to the baseline case, CFD results were used. Within this subsection the used numerical model is described.

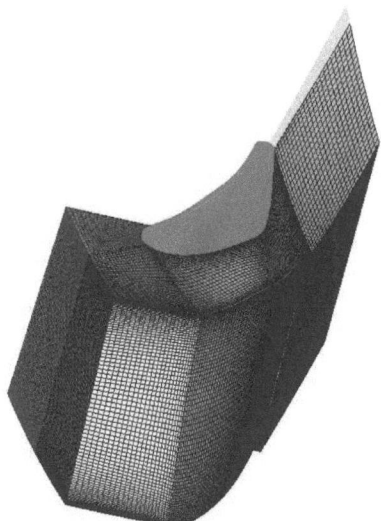

Figure 2.12: Multiblock Grid of the First Vane with Non-Axisymmetric Endwall

2.7 CFD

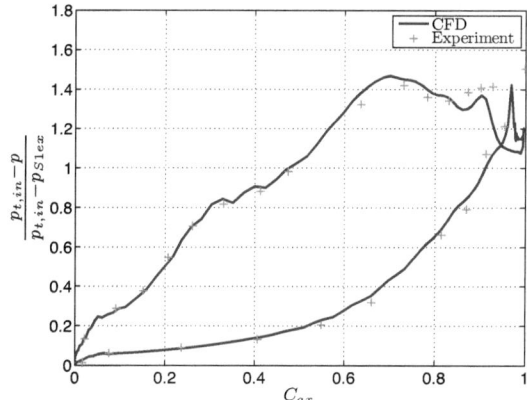

Figure 2.13: Second End Wall Design Vane Surface Pressure Distribution Comparison of Experiment and Simulation at 3% Span

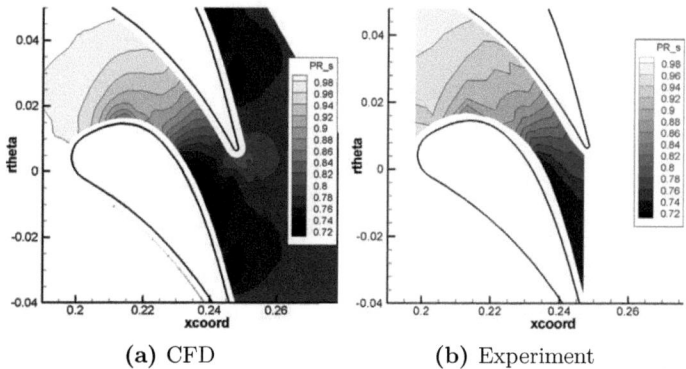

Figure 2.14: Second End Wall Design Vane Hub End Wall Pressure Field Comparison of Experiment and Simulation

The simulations were done with STAGE3D derived from the original Dawes code BT0B3D [?], running a discretization scheme from Jameson and Baker [?]. The simulation is a single row steady computation. The grid was generated with the in-house grid-generator MESHBOUND. MESHBOUND produces a multi-block grid consisting of an H-block, an up- and downstream wake-block and an O-block around the vane as seen in Figure 2.12. For further information refer to the work of Mischo [?]. In the course of this work the capabilities of the grid generator have been extended to include non-axisymmetric end walls and fillets. The inlet boundary conditions are taken from the measurements as shown in section 3.2. For the simulation the Baldwin Lomax turbulence model is chosen. In order to validate the model the experimental data is compared to the numerical results as shown in Figure 2.13 and Figure 2.14.

The pressure distribution at 3% span for second end wall design of the first vane shows a very good agreement between CFD and experiment. At the other measured span positions a similar level of agreement is seen. There is only some discrepancy at the late suction side as well as at the trailing edge. The trailing edge is however, a highly unsteady region, which can not be properly resolved by a steady CFD calculation nor by the pneumatic pressure measurement.

Also the second design vane hub end wall pressure field is properly predicted by the CFD model. This validation justifies a comparison between the experimental data of the non-axisymmetric end wall cases with the CFD data of the baseline axisymmetric case.

2.7.2 Time-Resolved CFD

In order to study the purge flow effects inside the blade passage a detailed numerical study was needed for all three different end wall configurations at different injection rates. In the following paragraphs the important information concerning these calculations are given.

Grid and Boundary Conditions The grid used for the time-resolved simulations is the same one as used during the design phase. It is a high quality multiblock grid (G3DMESH). As the ratio between vanes and blades is two to three two vane passages of the first and second vane row as well as three rotor passages are represented with periodic boundary conditions in the circumferential direction. In order to have a realistic rim seal flow field the physical cavity space as seen in Figure 2.3 underneath the first vane row is fully discretized with an interface to the first vane row hub end wall. A meridional cavity grid plane is shown in Figure 2.15. The total number of grid nodes is 10.8 million nodes. The non-dimensionalized wall distances on the airfoils and the end walls are about $y+ = 1$. At the inlet to the domain a constant total pressure of $1.4 bar$ and a constant total temperature

2.7 CFD

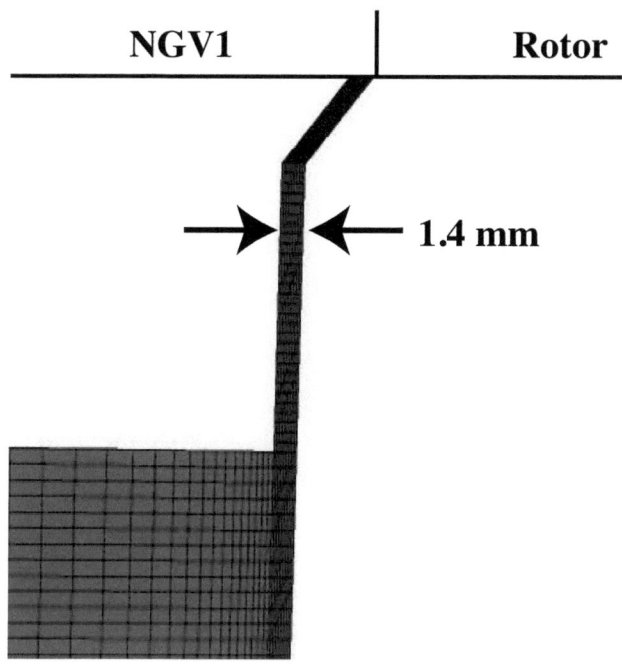

Figure 2.15: Meridional Grid Plane of Cavity

of $55°C$ were applied. At the exit the measured massflow for these inlet conditions was set. The cavity inlet was either modeled as a wall or with an inlet massflow of 0.9% of the main massflow. The cavity fluid temperature was set to $50°C$ as measured in the rig by a resistance thermometer.

Solver The time-resolved results were achieved with the commercial ANSYS CFX V11.0 software package. A steady run was done to derive the initial conditions for the unsteady simulation. The temporal resolution is 20 steps per blade passing event. The code is an implicit one and there are coefficient iterations for each step. This yielded a $0.33°$ shift of the rotor per time-step. For this simulation the $k - \epsilon$ turbulence model was used. The maximum residuals were found to be in the order of 10^{-3}, while the mass imbalances were in the order of 10^{-5}. The periodic convergence of the the unsteady simulations was judged based on the correlation coefficient of two pressure monitoring points at exit of the rotor row. Two consecutive

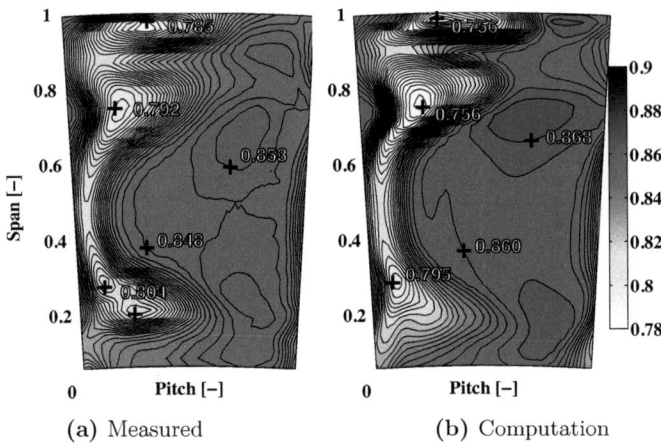

Figure 2.16: Comparison of Non-Dimensional Relative Total Pressure Pt_trel for Computation and Experiment at Rotor Exit $IR = -0.1\%$ for the Baseline End Wall

vane passage pressure events had to reach a correlation coefficient of above 99%.

Validation In order to validate the computational model the time-averaged results of the calculation are compared to the experimental data. The validation is done for the case with $IR = -0.1\%$ and $IR = 0.9\%$. As an example the normalized relative total pressure at rotor exit time-averaged in the relative frame is shown in Figure 2.16 and Figure 2.17. Generally Figure 2.16 and Figure 2.17 show the three zones of low relative total pressure caused by the hub and tip secondary flow as well as by the tip leakage vortex. The shape and radial position of the loss cores is well predicted. Figure 2.16 shows a good qualitative agreement between computation and experiment, which is also confirmed by the quantitative comparison as shown in Figure 2.18 (a). Especially the radial position of the loss cores is well captured by the computation.

When purge flow is applied the loss core at the hub is growing considerably as a consequence of the injection as seen in Figure 2.17. The hub loss core is radially further out compared to the $IR = -0.1\%$ case. All these trends are captured in the computational model.

2.7 CFD

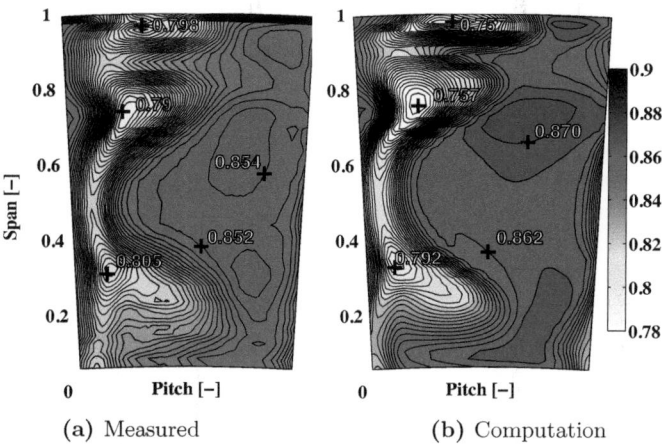

(a) Measured (b) Computation

Figure 2.17: Comparison of Non-Dimensional Relative Total Pressure Pt_trel for Computation and Experiment at Rotor Exit $IR = 0.9\%$ for the Baseline End Wall

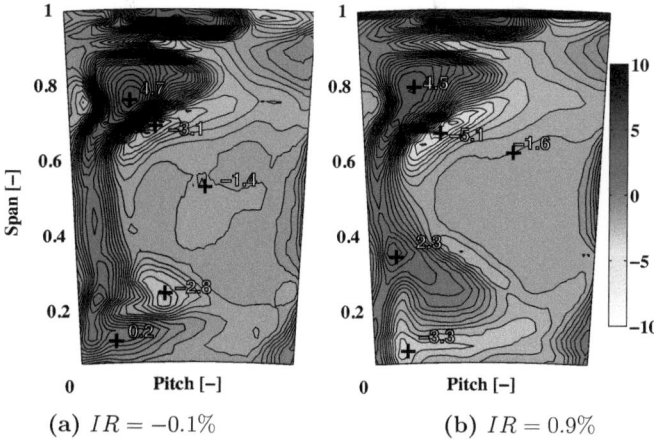

(a) $IR = -0.1\%$ (b) $IR = 0.9\%$

Figure 2.18: Calculated Relative Error of Computation for the Baseline End Wall

The relative error for both operating conditions is given in Figure 2.18. From the two Figures the following error levels can be extracted. The maximum error contour in the tip leakage and tip passage vortex is below 5% with and without leakage. In the wake region the mismatch is below 2%. In the hub passage vortex region the maximal error is within 1% without injection and within ±3% with injection. In the free-stream region the computation over-predicts the relative total pressure by about 1%. If the mass-averaged error is calculated the error magnitude is within 2%.

3 Non-Axisymmetric End Wall Profiling

One of the main objective of this work is the investigation of the effects of non- axisymmetric end wall profiling in an axial turbine. So far a number of investigations have been executed in linear cascades. However, only very limited time-resolved data is available for the flow physics of rotating turbine stages equipped with non-axisymmetric profiling. In the course of this work two different designs were investigated (1. Gen. and 2. Gen.). Only the first stage of the 1.5-stage turbine was equipped with profiled end walls. This implies a tip and hub end wall design for the first nozzle guide vane row and a hub end wall design for the unshrouded rotor. The investigation offers some details on the unsteady effects of non-axisymmetric end wall profiling. At this stage the influence of purge flow is neglected. Therefore, all measurements presented in this chapter were measured with small suction of 0.1% of the main massflow.

3.1 Design Methodology

The end wall design was executed in a cooperative program with MTU Aero Engines. For further details refer to Germain et al.[?]. The main features of the methodology are the following:

- The end wall parameterization consists of a combination of various objects ("bumps"), each represented by an axial and a circumferential function, allowing complex shapes. See Figure 3.1. This parameterization was initially introduced by Nagel et al. [?]. However, the underlying mathematical functions have been modified in order to have more control over more complex shapes.

- The CFD mesh is created through a fully integrated CAD / CFD-meshing system, which makes use of a high quality multiblock grid

(G3DMESH [?]). The CFD grid shows non-dimensionalized wall distances on the airfoils and the end walls of about $y+ = 1$. The resulting number of nodes is about 1.5 million per aerodynamic row. The tip clearance gap is fully discretized using 17 points radially.

- The MTU in-house optimization system is used to close the loop of geometry definition, meshing, flow solving and post-processing by the sequential quadratic programming optimization algorithm donlp2 [?], which allows the solution of non-linear constrained problems.

- The primary objective of the optimization is to reduce the secondary kinetic energy. A second order target is to improve efficiency. The design space was constrained by keeping a constant averaged outflow swirl angle and therefore, the same capacity. In order to correct possible capacity changes between the non-axisymmetric end wall cases and the baseline case the first vane was skewed. For the first contoured geometry the stator is skewed by 0.55°, thereby maintaining the capacity. For the second design the skewing of the airfoil to maintain the same capacity is $-0.17°$.

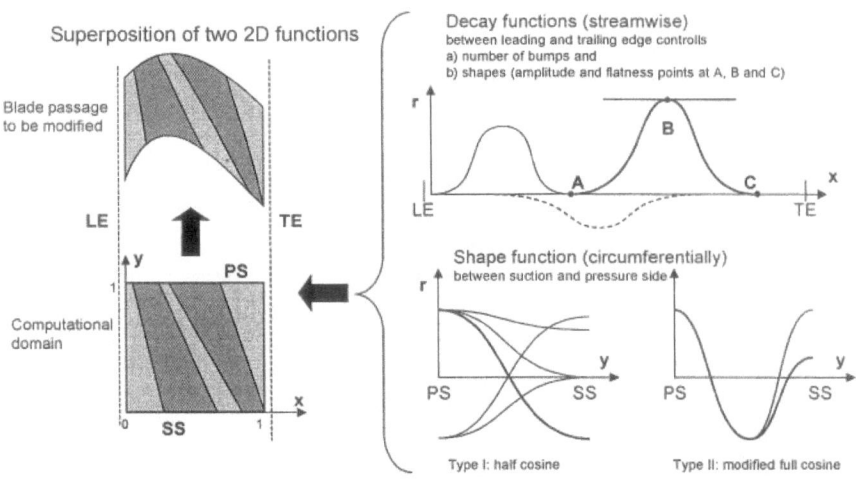

Figure 3.1: Parameterization of the Non-Axisymmetric End Walls [?]

The design calculations are performed using the URANS code TRACE developed at DLR and MTU, which has been specially designed for the

3.1 Design Methodology

simulation of steady and unsteady turbomachinery flows. The turbulence closure is modelled by the Wilcox $k - \omega$ two-equation model in a low-Reynolds version with compressibility extension. The boundary layers are computed following the low-Reynolds approach. A transition model is used on the airfoil suction side and pressure side, based on the modified correlations of Abu-Ghannam and Shaw [?]. The model can be activated on the airfoil, while the end wall boundary layers are assumed to be of a turbulent nature.

In the second design phase, the geometry definition system (parameterization and grid generation) as well as the flow solver remained the same. However, some new design methodology features have been integrated for the second design phase as well as some new design constraints:

- The aim was to receive a more homogenous radial swirl angle distribution out of the first vane row. It is expected that this has a positive effect on the behaviour of the rotor compared to the first design exercise.

- In a new design approach the hub and tip end wall are first optimized independently. This ensures that the gradient algorithm is equally weighting each end wall in the initial optimization. In a second step both end walls are further optimized together.

- The fillet radii geometry is represented by the design meshes.

As shown in Figures 3.2 (a & d) the first vane hub end walls of both non-axisymmetric designs are characterized by a suction side dent. However, the second design has a smaller amplitude than the first design. Furthermore, the second design has additionally a characteristic ridge at the trailing edge. This geometrical feature as well as the reduced amplitude of the second design are expected to reduce the hub overturning. The vane tip end wall of the first design is characterized by a depression on the suction and pressure side with a stronger amplitude on the suction side (Figure 3.2 (b)). The second design has a more "classic" shape with an elevation along the pressure side and a depression along the suction side (Figure 3.2 (e)). The profound change in shape of the vane tip end wall between the two end wall designs is expected to be the result of the independent optimization of both vane end walls during the second design phase. During the first phase both end walls were optimized together. The rotor hub end wall of the first design shows an bump on the early pressure side and a dent shortly before the throat plane (Figure 3.2 (c)). The second design rotor hub (Figure 3.2 (f)) shows larger amplitudes and a larger extend, while the suction side trough moves towards the leading edge.

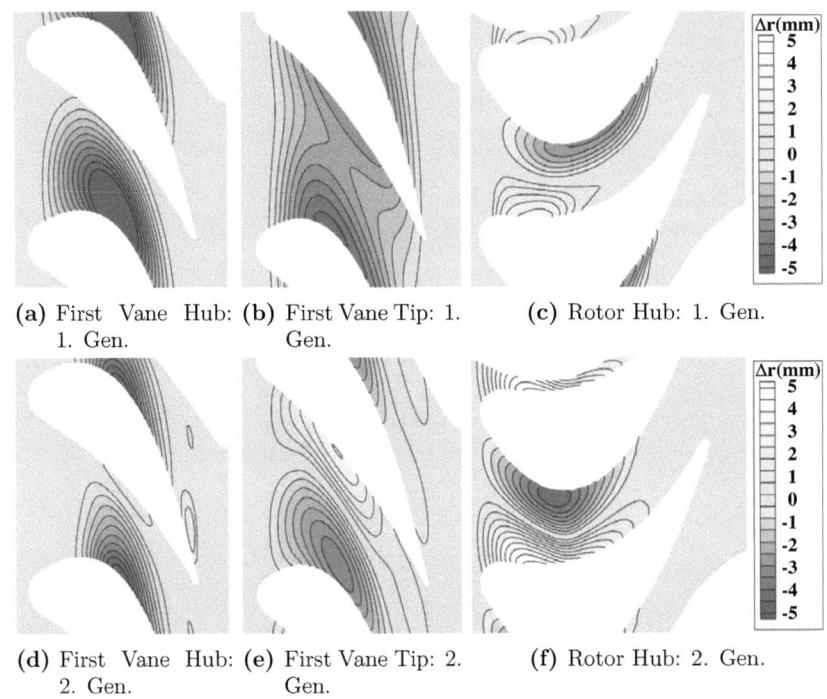

(a) First Vane Hub: 1. Gen.
(b) First Vane Tip: 1. Gen.
(c) Rotor Hub: 1. Gen.
(d) First Vane Hub: 2. Gen.
(e) First Vane Tip: 2. Gen.
(f) Rotor Hub: 2. Gen.

Figure 3.2: Non-Axisymmetric End Wall Shapes from Optimization

To get an idea of the three dimensional shape of non-axisymmetric end wall profiling an illustration of the vane hub end wall of the first design is given in Figure 3.3.

3.2 Operating and Inlet Conditions of the Turbine

During measurement the turbine 1.5-stage total-to-static pressure ratio is kept constant at $\Pi_{1.5} = 1.65$. The entry temperature is kept constant to

3.2 Operating and Inlet Conditions of the Turbine

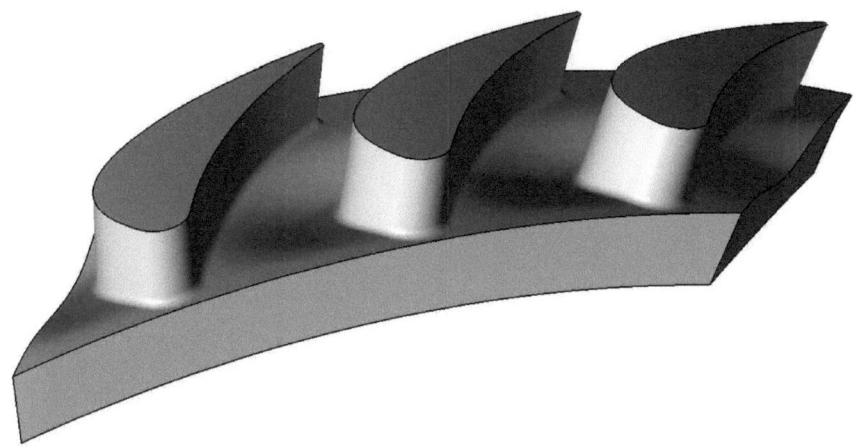

Figure 3.3: Three Dimensional View of the 1. Gen. First Vane Hub End Wall

permit an accurate comparison between measurements made on different days. To account for the change in ambient pressure on different measurement days the pressures are non-dimensionalized by the respective turbine inlet total pressure. Table 3.1 shows the design point operating conditions. All measurements have been executed at the design rotational speed of $2700 rpm$.

Pressure Ratio $\Pi_{1.5}$	$1.65 \pm 0.4\%$	$[-]$
Inlet Total Temperature $T_{t,in}$	328 ± 0.2	$[K]$
Capacity $\left(\dot{m}\sqrt{T_{t,in}}\right)/p_{t,in}$	$152 \pm 0.2\%$	$\left[\frac{kg \cdot K^{1/2}}{s \cdot bar}\right]$
Dimensionless Rotational Speed $N/\sqrt{T_{t,in}}$	2.48 ± 0.05	$\left[\frac{r.p.s.}{K^{1/2}}\right]$
Loading Coefficient $\Delta h/U^2$	2.36	$[-]$
Flow Coefficient C_x/U	0.59	$[-]$

Table 3.1: Design Point Operating Conditions

In Figure 3.4 the inlet total pressure profile and the yaw angle profile for all three geometries are given. The inlet traverse plane is 50.4% of the first vane axial chord upstream of the first vane leading edge and covers one

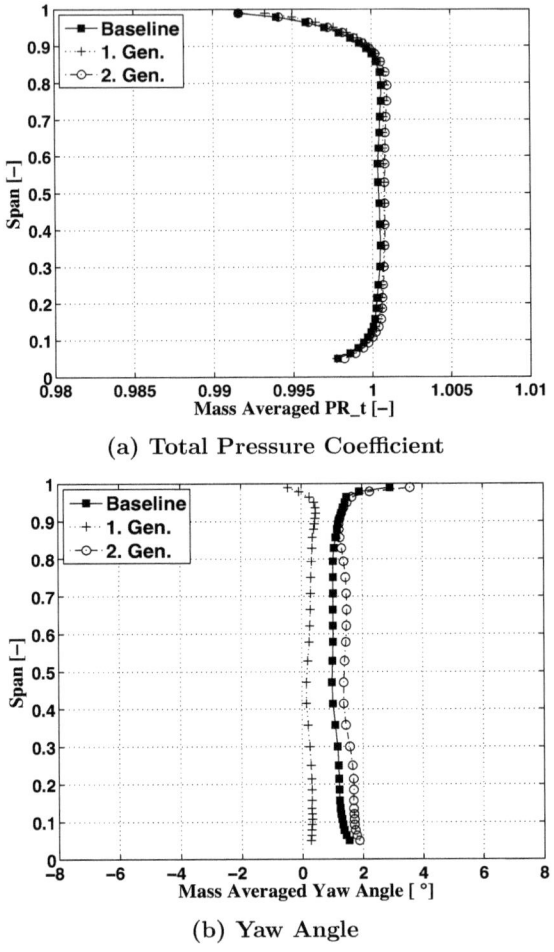

Figure 3.4: Measured Mass-Averaged Inlet Conditions

stator pitch. It starts at -90% pitch relative to the trailing edge of the first vane. The inlet flow is close to axial and the variation at mid-height is within $\pm 0.6°$, which is in the order of the measurement accuracy. The inlet total pressure variation in the free-stream is 2% of the inlet dynamic head. The free-stream turbulence intensity is around 2.5%. The boundary layer is quite well defined by a power law profile as defined in Equation 3.1

3.3 Total-to-Total Stage Efficiency

with $\delta = 10mm$ and $n = 10$.

$$\frac{C}{C_\infty} = \left(\frac{y}{\delta}\right)^{1/n} \tag{3.1}$$

3.3 Total-to-Total Stage Efficiency

$$\eta_{tt} = \frac{\frac{\omega \cdot M}{\dot{m}_v \cdot c_p \cdot T_{t,in}}}{1 - \left(1 - \frac{IR}{100}\right) \cdot \left(\frac{p_{t,R1ex}}{p_{t,in}}\right)^{\frac{\gamma-1}{\gamma}} - \frac{IR}{100} \cdot \left(\frac{p_{t,R1ex}}{p_{t,cav}}\right)^{\frac{\gamma-1}{\gamma}}} \tag{3.2}$$

Table 3.2 gives the absolute efficiency improvements based on Equation 3.2. In the numerator of Equation 3.2 the power coefficient is given. The power coefficient is the turbine power as the product of rotational speed ω and torque M divided by the product of the main massflow $\dot{m}_v$ and the inlet total enthalpy $c_p \cdot T_{t,in}$. In the denominator the isentropic enthalpy changes of the main flow and the purge flow can be found. The first design shows an improvement in total-to-total efficiency of 1%±0.32% relative to a baseline efficiency of 91% ± 0.37%. The efficiency benefit with the second end wall design is only one third of the value achieved with the first design.

Variable	Baseline.	1. Gen.	2. Gen.
η_{tt}	91.0% ± 0.37%	92.0% ± 0.37%	91.3% ± 0.37%
$\Delta\eta_{tt}$		Δ1.0% ± 0.32%	Δ0.3% ± 0.32%

Table 3.2: Absolute Total-to-Total Efficiency Improvements

3.4 First Vane

3.4.1 End Wall Pressure Field

In this section, the end wall pressure data is presented in order to quantify the effect of end wall profiling. As no experimental data is available for the baseline case, CFD results are used. This approach is justified by a close agreement between CFD and experimental results for the second profiled end wall as presented in subsection 2.7. The end wall pressure measurement has a resolution of 8 points in the circumferential direction and 12 points in the streamwise direction. The exact distribution can be found in Figure 2.4. Figures 3.5, 3.6 and 3.7 show the the first vane pressure field of the hub and tip end wall for all three geometries. Generally, the end wall contouring results in a reduced cross passage pressure gradient in the late passage (light arrows) and a promoted crossflow in the early passage (dark arrows). This potentially reduces the cross-flow in the late passage. At the hub suction side of the second end wall geometry at $0.215m$ a pressure minimum can be found. This results in a increased skewing of the boundary layer in the early passage. Furthermore, the late suction peak value is reduced with the second end wall design. At the tip the first design shows an early suction side minimum similar to the one of the second design at the hub. Furthermore, the first end wall contour shows a very strong cross passage pressure gradient at around $0.21m$. Finally, the first design shows a reduced late suction side peak. Generally the profiling at the hub is more effective in suppressing the late passage cross passage pressure gradient than the profiling at the tip.

3.4.2 Lift Plots

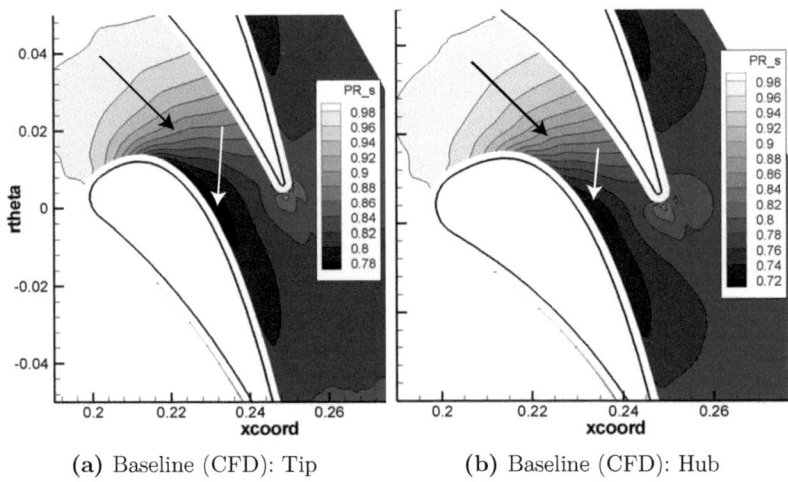

(a) Baseline (CFD): Tip   (b) Baseline (CFD): Hub

Figure 3.5: Predicted Stator 1 End Wall Pressure Field PR_s (Basline Geometry)

3.4 First Vane

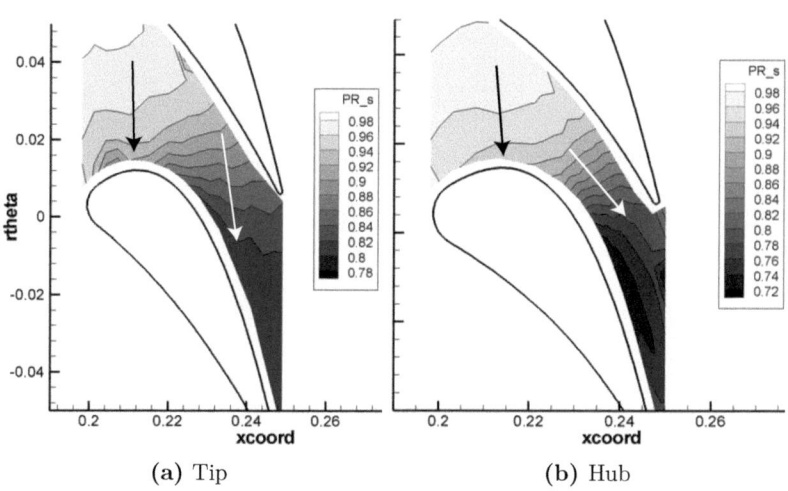

(a) Tip (b) Hub

Figure 3.6: Measured Stator 1 End Wall Pressure Field PR_s (1. Gen. End Wall)

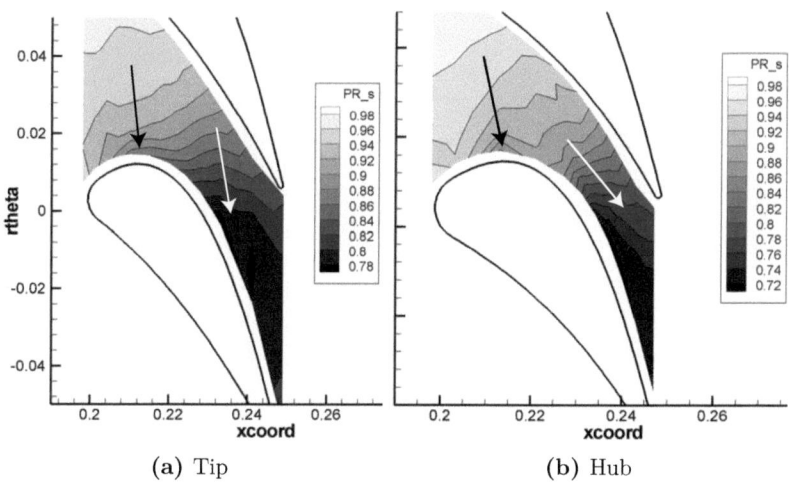

(a) Tip (b) Hub

Figure 3.7: Measured Stator 1 End Wall Pressure Field PR_s (2. Gen. End Wall)

3 Non-Axisymmetric End Wall Profiling

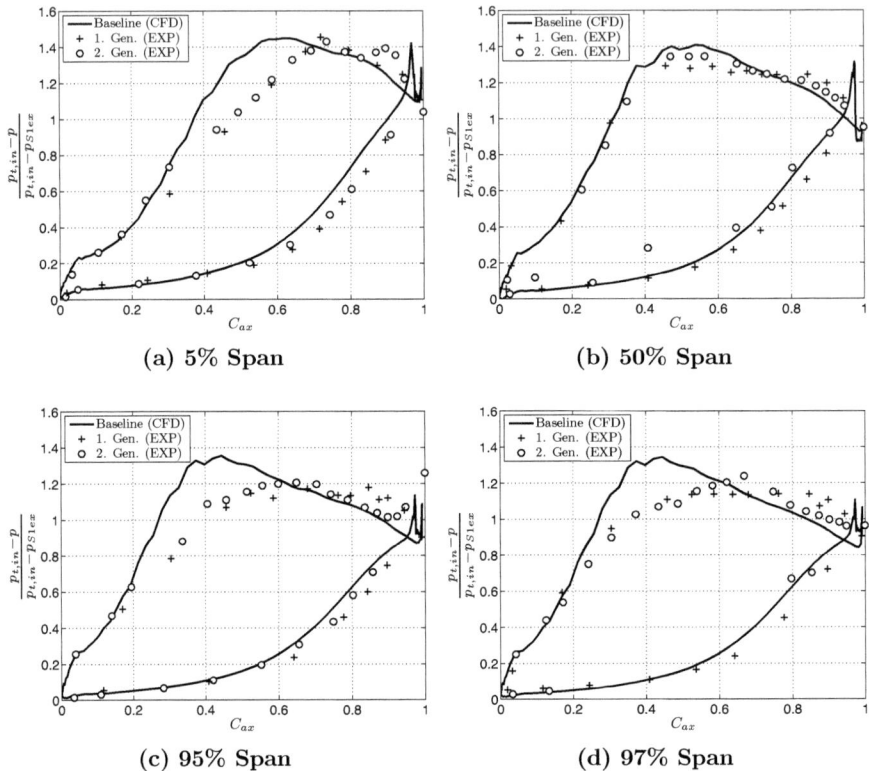

Figure 3.8: Lift Plots for First Vane

Figure 3.8 shows the lift plots of the first vane for all three geometries at a hub position (5% span), at mid-height (50% span) and at two tip positions (95% & 97% span). As there is also no vane surface pressure measurement installed in the axisymmetric case the non-axisymmetric cases have to be again compared to CFD results of the baseline case. The comparison of measurement and calculation for the second end wall design can be seen again in section 2.7. The span position is given as percentage value of the axisymmetric geometry. At the hub (5% span) the non-axisymmetric profiling creates a more aft-loaded pressure profile. The second design shows a reacceleration on the late suction side, which is most probably the result of the characteristic ridge near the trailing edge. It has been introduced to

3.4 First Vane

reduce the overturning at the hub end wall. At mid-height up to 90% of the axial chord C_{ax} the first design shows very little diffusion compared to the baseline. At the tip end wall (95% span) the non-axisymmetric cases show a more aft-loaded pressure distribution. The peak suction value is reduced with end wall profiling by 15% of the ideal exit dynamic head resulting in an off-loading of the tip. This is the result of reduced blockage due the profiling. With the first design the late suction side shows almost no diffusion and shows a reacceleration at 85% axial chord C_{ax}. At the 97% span position the profiled cases show once more a lower suction peak. The second end wall doesn't show any measurement points on the pressure side as a result of the stronger pressure side profile. With the second design the flow is accelerated up to 67% axial chord. With the first design the acceleration ends at 45% axial chord. Additionally the suction peak with the second design is 9% higher. These effects finally result in a much stronger diffusion on the late suction side of the second design case.

3.4.3 Vane Suction Side Pressure Field (CFD)

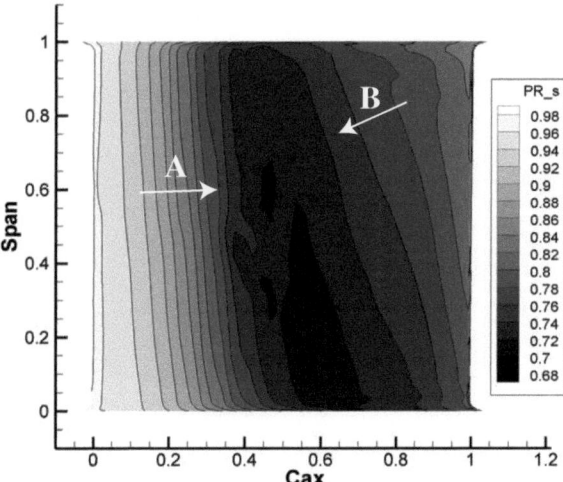

Figure 3.9: Predicted Pressure Field on First Vane Suction Side PR_s (Baseline Geometry)

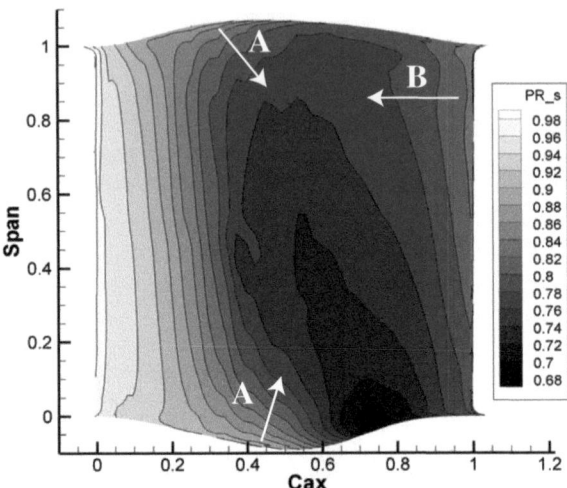

Figure 3.10: Predicted Pressure Field on First Vane Suction Side PR_s (1. Gen. End Wall)

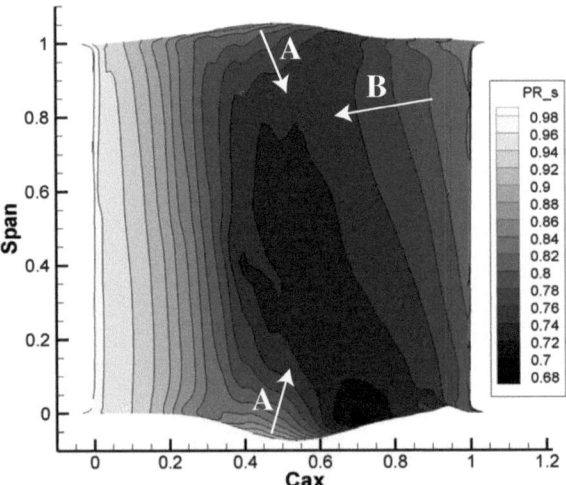

Figure 3.11: Predicted Pressure Field on First Vane Suction Side PR_s (2. Gen. End Wall)

3.4 First Vane

The pressure measurements in the non-axisymmetric cases at 5%, 50%, 95% and 97% span are not sufficient to get a global picture of the first vane suction side pressure field. Therefore, Figures 3.9, 3.10 and 3.11 present the computed suction side pressure field. However, the computed pressure shows a good level of agreement with the available measurements as seen for example in Figure 2.13. Figures 3.10 and 3.11 show also nicely the streamline curvature due to the profiling. It can be seen that the second design is introducing stronger concave curvature at the hub as well as at the tip end wall. At the tip of the second design this results in more convex curvature towards the trailing edge. Furthermore, the second end wall design is characterized by a very aggressive convex curvature at the late suction side hub corner which is introduced by the ridge. However, towards the rim there is again a part of concave curvature. With this the circumferential pressure gradient at the rim is reduced. The early passage suction side pressure field (up to 60% axial chord) of the baseline case doesn't show a strong radial pressure gradients apart of some minor end wall effects **A**. In the tip region of the late suction side the baseline pressure field is characterized by a radial tip to hub pressure gradient. As a consequence there is radial inward migration **B**. On the early suction side the profiling in contrast generates a strong pressure gradient away from both the hub and the tip end wall towards the annulus center **A**. The effect of the second design is even stronger as a result of stronger curvature. However, the strong convex curvature of the second design towards the trailing edge creates the strong diffusion as seen in Figure 3.8 (d). The tip to hub pressure gradient on the late suction side is reduced most with the first design.

3.4.4 First Vane Intrarow Measurement

In order to study the development of the tip secondary flow structure the measured total pressure coefficient at 66% and 90% axial chord is given in Figure 3.12. At 66% axial chord the first end wall design shows a larger zone of total pressure loss. Additionally the peak total pressure loss is slightly higher with the first design at this axial location. At 90% axial chord the profiling has been smoothed out. Therefore, no influence of curvature beyond this point can be expected. At this plane the covered area of the tip loss core is about twice as big with the second design as with the first design. Therefore, from 66% axial chord to 90% axial chord the second design has to cause much more total pressure loss. The influence of the strong diffusion

3 Non-Axisymmetric End Wall Profiling

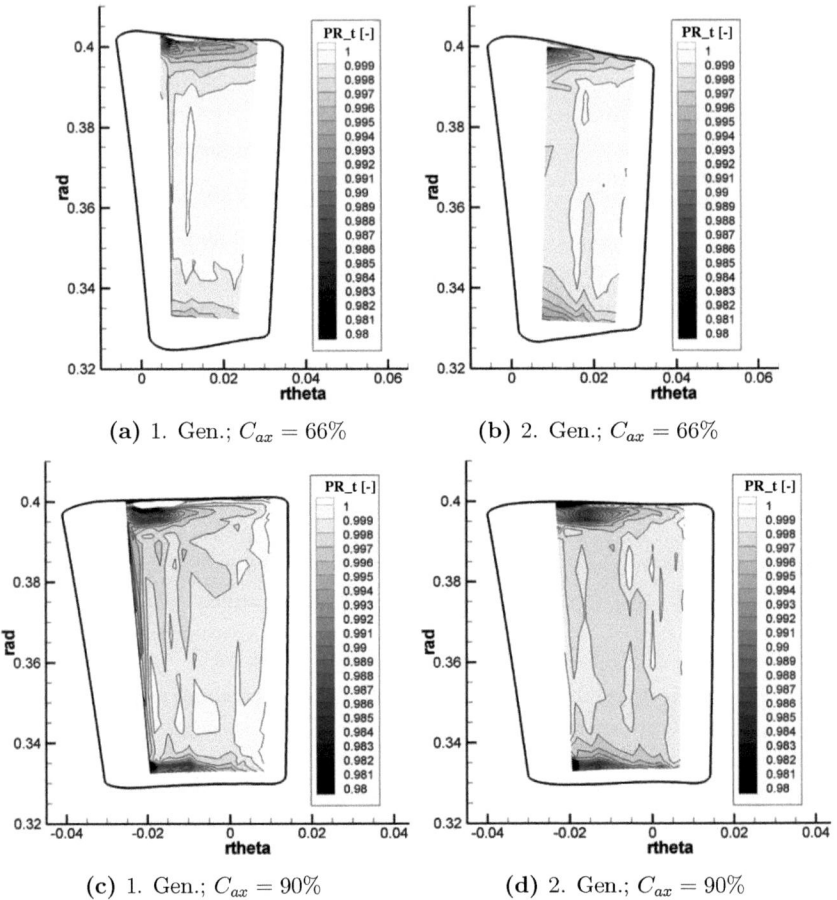

Figure 3.12: Measured Total Pressure Coefficient PR_t at Two Circumferential Planes Inside S1

caused by the strong second design late suction side curvature is most pronounced within these two planes. Therefore, it is assumed that this additional loss is the result of some sort of separation (corner stall) caused by the strong late suction side diffusion.

3.4 First Vane

3.4.5 First Vane Row Exit Flowfield

In this subsection the measured effect of the profiling at exit of the first vane row is shown.

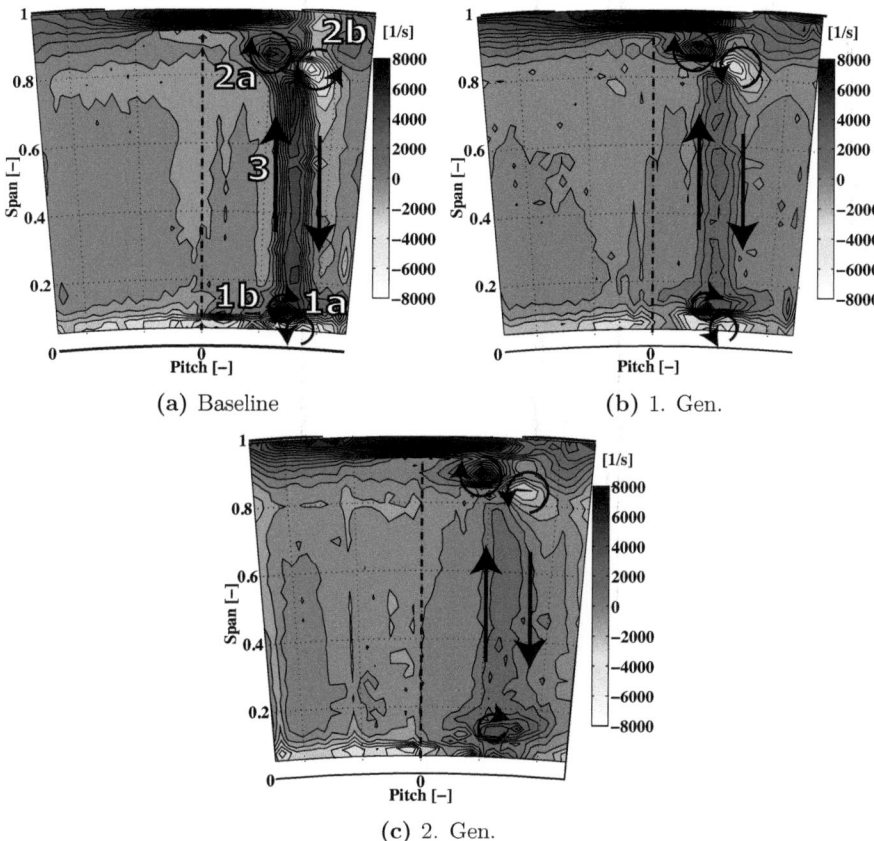

Figure 3.13: Measured Streamwise Vorticity ω_s at S1ex

Figure 3.13 shows the time-averaged streamwise vorticity plot at the exit of the first vane row. The tip secondary flow structures can be found around 80% span. The zone of positive vorticity is the tip passage vortex and is labelled **2a**. This vortex becomes smaller and more intense with both non-axisymmetric end walls. The zone of negative vorticity at the tip **2b** is the tip trailing shed vortex. This vortex is increased with the first design and reduced with the second design relative to the baseline case. In contrast to the two tip secondary vortices which are next to each other the hub secondary structures are on top of each other. The hub trailing shed vortex as zone of positive vorticity can be found at 10% span and is labelled **1b**. This vortex is reduced by both non-axisymmetric end wall designs relative to the baseline case. There is as well a stronger reduction than at the tip. The hub passage vortex **1a** is the zone of negative vorticity underneath the hub trailing shed vortex. This zone is absent in the second end wall design case. Eventually the vortex is now just so close to the hub end wall that it isn't captured by the probe. However, the main reduction of vorticity is seen between 20% and 75% span **3**. This band of positive vorticity is trailing shed vorticity which is the result of strong inward and outward migration of the wake fluid as indicated by the arrows. Both non-axisymmetric end walls show a reduction by a factor two in this region.

In Figure 3.14 the absolute frame secondary kinetic energy contours at the first vane exit are plotted. There are two zones of high secondary kinetic energy. One associated with the tip secondary flow structures covering a radial region between the tip and 70% span. The second is the result of the hub secondary flow structures below 20% span. However, only the top part of the hub passage vortex is captured by the measurement. Generally the tip secondary flow structures cause more secondary kinetic energy. It can also be seen that the secondary kinetic energy is high on the periphery of the vortices and not in the vortex center. At exit of the first vane row predominately the circumferential secondary velocity component is contributing to the secondary kinetic energy. In the largest and strongest secondary kinetic energy zone between 80% and 90% span it is difficult to see a distinct difference between the three cases. The smaller tip secondary kinetic energy region between 90% span and the tip end wall is smaller with the first end wall design compared to the baseline end wall, but more intense. The second end wall geometry gives a smaller region which is also less intense compared to the baseline geometry. At the hub the region of highest secondary kinetic energy is at 10% span caused by both the hub trailing shed and the hub passage vortex. Both profiled cases show a reduction in covered area. At about 17% span there is the high secondary kinetic energy zone caused by the hub trailing shed vortex. In this region the first end wall design is showing the lowest value. However, the trailing shed vorticity as the dominant flow structure in the baseline case is not

3.4 First Vane

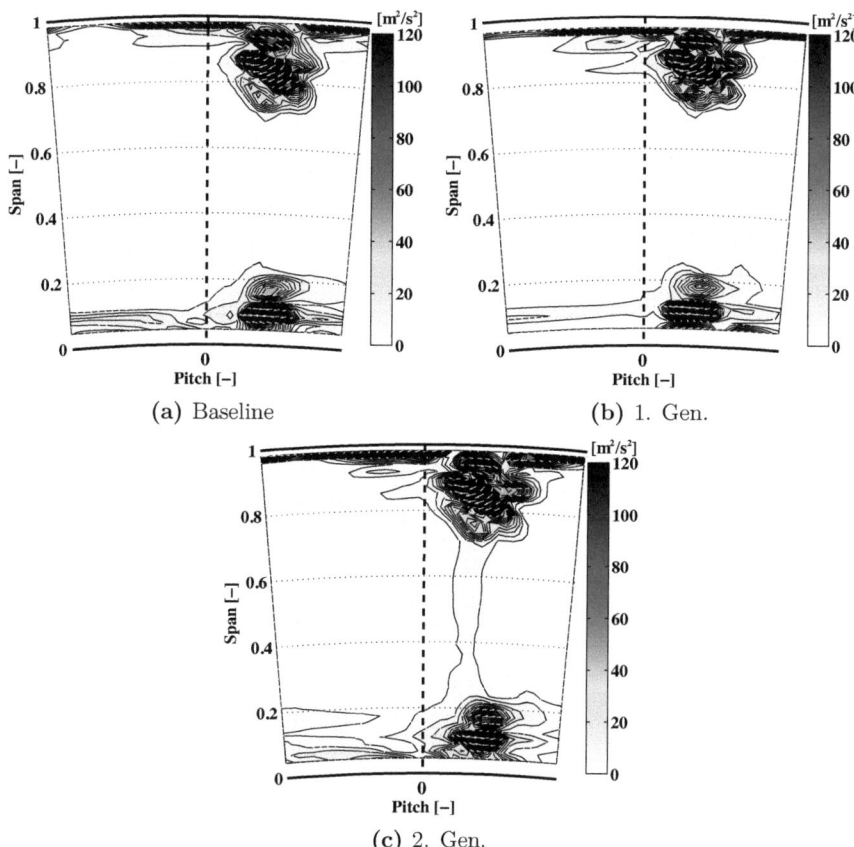

Figure 3.14: Measured Secondary Kinetic Energy SKE (see section 2.5.3) at S1ex

represented by secondary kinetic energy. Therefore, it is impossible to predict the effect of profiling on this structure in terms of secondary kinetic energy only.

The consequences of the profiling in terms of loss are shown in Figure 3.15 covering one stator pitch. The hub **1** and tip **2** secondary losses can be found around 10% and 80% span respectively. Additionally the identified vortices from Figure 3.13 are drawn. The secondary loss cores are connected by a band of high loss, which represents the vane wake **3**. At the hub **1** both end

3 Non-Axisymmetric End Wall Profiling

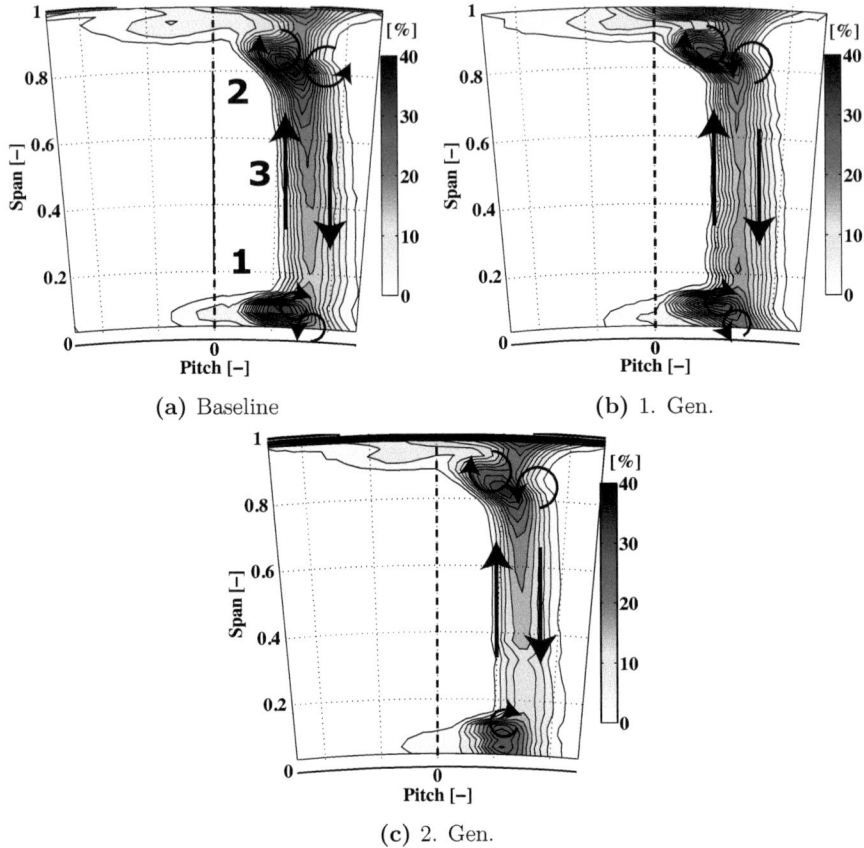

Figure 3.15: Measured Steady Loss Coefficient Y (see Table 2.6) at S1ex

wall designs show a reduction of loss relative to the baseline case which is the result of reduced secondary flow as seen in terms of streamwise vorticity and secondary kinetic energy reductions. The second hub end wall design results in a changed flow structure shape with a reduced circumferential extent. Furthermore the second design case shows two loss cores instead of one. It is possible that the characteristic ridge of the second design is creating an additional flow structure. There is only a small reduction of loss with profiling for the tip secondary flow structure **2**. This is the result of only small streamwise vorticity and secondary kinetic energy reductions in

3.4 First Vane

this region. However, the largest differences in loss are found outside of the secondary flow zones. The second end wall design shows the highest loss towards the casing which is probably the result of a separation as a result of to much diffusion on the late suction side. Both non-axisymmetric end wall profiles reduce loss in the outer wake region. Most probably this is the result of a reduction of the suction side peak velocities and the associated profile loss. As seen in the lift plots Figure 3.8 the profiling results in a reduced suction side peak and therefore lower velocities on the suction side. The reason for this is an alleviation of vane blockage as the profiling adds area in the throat region. Furthermore the profiling results in an aft-loaded profile which potentially delays transition resulting in a further profile loss reduction.

In order to quantify the loss reductions with profiling the circumferentially mass-averaged total pressure loss coefficient at exit of the first vane row is plotted in Figure 3.16 (a). The second design is characterized by the highest loss in the region from 80% span towards the tip which is probably the result of corner stall caused by a too aggressive late suction side diffusion. In the tip secondary flow region at 80% span the effect of end wall profiling is small. Both end wall designs reduce the wake loss as a result of reduced peak Mach numbers on the suction side and maybe due to a delayed transition as a result of aft-loading. The first design shows the lowest loss between 30% and 80% span and the second design shows the lowest loss from 30% span down to the hub. The hub secondary loss is reduced by both designs. The second design shows the lowest loss at 10% span but more at the last measurement point. Generally the second design results in more radial spread of the hub secondary loss resulting from the two loss cores as observed previously. Additionally it has to be stated that the last 4% of the flow towards the hub are not captured by the probe. This means about 3% in mass flow. This is a region of high loss fluid. Therefore, the available data doesn't allow to make any statements on the nature of the hub end wall boundary layer and the associated changes. However, this will result in a under-prediction of the mass-averaged first vane loss coefficient since high loss fluid is not considered. An assessment based the computational results reveals a 8% under-prediction.

Figure 3.16 (b) shows the circumferentially mass-averaged deviation at exit of the first vane row in order to comment on the effect of the secondary flows. At the casing the overturning introduced by the tip passage vortex can be found. Between 70% and 95% span the fluid is underturned by

74　　　　　　　　　　　　3 Non-Axisymmetric End Wall Profiling

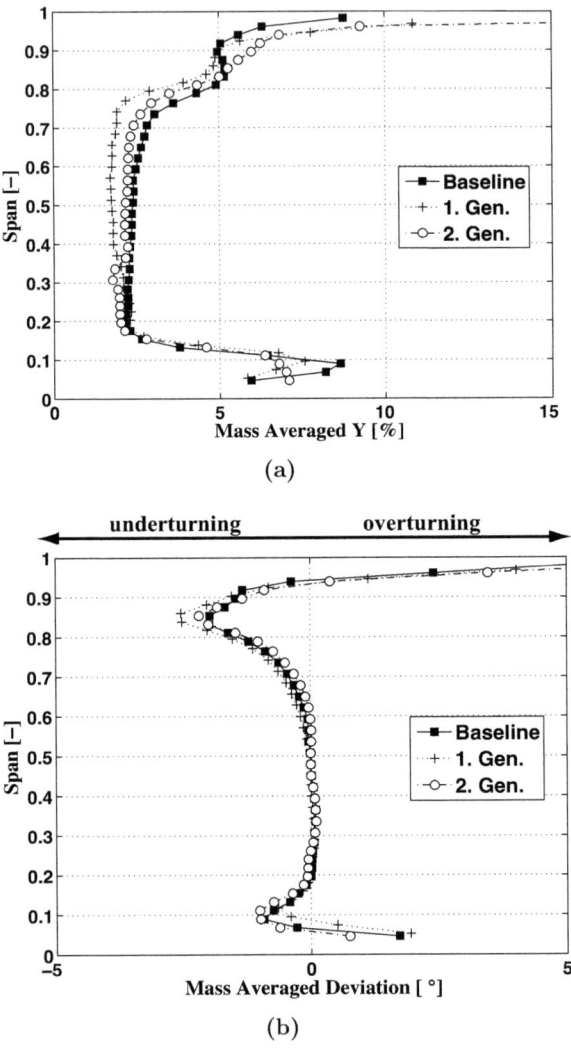

Figure 3.16: Measured Mass-Averaged Data at S1ex

the tip passage vortex. The tip passage vortex of the baseline geometry introduces the lowest underturning. With the second end wall design the underturning is increased. With the first end wall the underturning is

3.4 First Vane

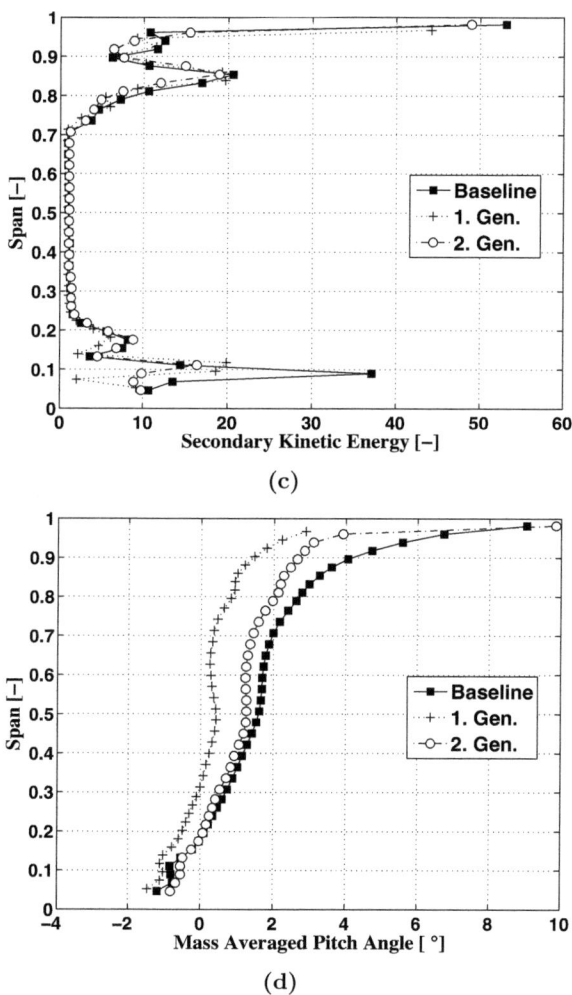

Figure 3.16: Measured Mass-Averaged Data at S1ex

increased once more. Therefore, it is once more confirmed that the two tip profiles didn't reduce the secondary flow at the tip. They rather increased it. In between 20% and 5% span the underturned fluid of the hub passage vortex can be found. The underturning is only reduced with the first end wall design. The second end wall design and the baseline case show the same level of peak underturning. Towards the hub the overturning resolved by the measurement has been reduced by about 1° with the latest design compared to the other two cases as it was the design intention. Therefore, both hub end wall designs reduce the hub secondary effect meaning the difference between maximum under- and overturning. The second design at the same time also reduces the hub overturning as intended.

The circumferentially mass-averaged secondary kinetic energy at the exit of the first vane row as seen in Figure 3.16 (c) has been non-dimensionalized by the baseline mid-height value. The highest levels of secondary kinetic energy are found in the tip boundary layer. There are always two peaks associated with a secondary flow zone. At 90% and 85% span there are the two secondary kinetic peaks associated with the tip secondary flow structures. The first peak at 90% is reduced most with the second design. However, the loss at the same time is the highest. The next peak is at 85% span, which is associated with the tip passage vortex. The differences between the three cases are negligible as already seen in the contour plots. Therefore, there are also no large differences between the loss levels of the three cases in this region. In the wake region the secondary kinetic energy levels between the cases do not differ as the effects of the trailing shed vorticity are not captured. At the hub there are again two secondary kinetic energy peaks at 18% and 9% span respectively associated with the hub trailing shed and passage vortex. The upper peak of the baseline and second design are about the same, while the first design shows a slight reduction. In the lower peak the secondary kinetic energy level for both non-axisymmetric cases is halved. In this region the secondary kinetic energy reduction correlates nicely with the loss reduction.

Figure 3.16 (d) shows the circumferentially mass-averaged pitch angle at exit of the first vane row. In this representation the influence of the secondary flow is not seen as the circumferential gradient associated with streamwise vorticity is averaged out. Generally the wake fluid at the exit of a vane row is migrating radially in due to the pressure gradient resulting from the swirling flow. However, the pitch angle towards the tip in the baseline case and the second end wall design is positive, meaning pitching out. The trailing shed vorticity in the baseline case is generated as a result of the wake migrating strongly inwards and the suction side fluid outside the viscous region pitching outwards. The average pitch angle in the second case is probably showing that the flow is recovering after the separation.

3.4 First Vane

The separation creates blockage pushing the flow inwards. At vane row exit the flow recovers an moves radially out again.

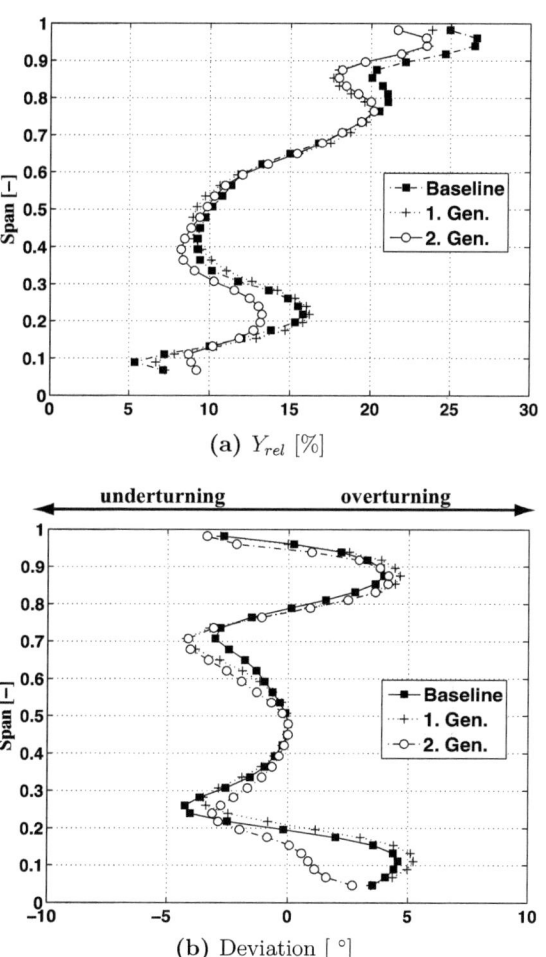

(a) Y_{rel} [%]

(b) Deviation [°]

Figure 3.17: Measured Mass-Averaged Data at R1ex

3.5 Rotor Exit Flowfield

In this section data from the traverse at the rotor exit is presented. As the rotor is unshrouded changes in the outer annulus are expected to be a consequence of the changed flowfield at the blade row inlet.

The radial distribution of the relative total pressure loss coefficient Y_{rel} as seen in Figure 3.17 (a) is determined with the mass-averaged rotor inlet and exit relative total pressure values from the same radial height. The rotor inlet relative total pressure between the three cases shows a maximal variation of below 0.5% of the turbine inlet value. The tip leakage vortex loss is located at 95% span, the tip passage vortex loss at around 80% and the hub passage vortex loss at 22%. While the difference between the baseline and the first design in the lower annulus is negligible, the second design shows a reduction of more than 2% of the peak loss. This results in an integral relative total pressure loss reduction of 6% for the second design rotor end wall relative to the baseline case. This is the result of the improved end wall geometry at the rotor hub and an improved first vane hub flowfield. The second end wall geometry keeps the low momentum or high loss fluid close to the end wall. Therefore, the second design shows the highest loss level for the measured points closest to the hub end wall. The first end wall design shows a integrated reduction of 3% relative to the baseline which is only the result of a reduction at the tip.

Figure 3.17 (b) shows the deviations from the mean relative flow angle for all three cases. At the casing there is an undertuned region caused by the tip leakage vortex. At 87.5% span the tip leakage vortex and the tip passage vortex both introduce overturning. At 70% span there is an underturned region due to the tip passage vortex. The profiling causes an enhanced vorticity at the tip and exit of the first vane row. As seen in subsection 3.6.1 the profiling leads to a lower reaction which in turn results in higher Mach numbers at rotor inlet. The higher Mach numbers, predominantly at the tip lead to positive incidence onto the rotor. Therefore, the turning at the rotor tip will be increased which enhances the tip secondary flows. Therefore, the overturning at 87.5% span as well as the underturning at 70% span are increased with profiling. Around 25% span the flow is underturned as a result of the hub secondary flow and around 10% span the flow is overturned by the hub secondary flow. With the first end wall geometry the underturning at the hub is decreased by 0.8°, while the overturning is increased by 0.7°. The maximum underturning is 2.1% of the span further

3.5 Rotor Exit Flowfield

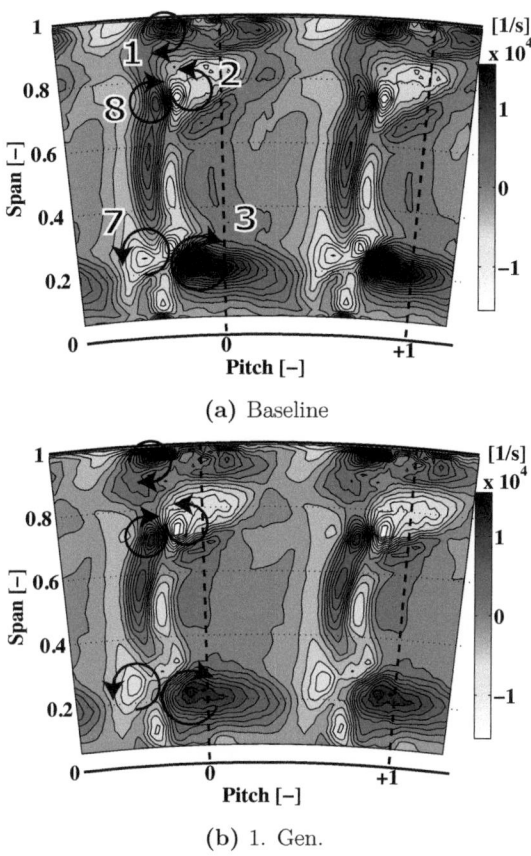

Figure 3.18: Measured Streamwise Vorticity ω_s at R1ex in the Rotor Relative Frame

out. With the second end wall geometry the underturning is decreased by 1.1° and 2.1% of the span closer to the hub compared to the baseline case. There is no pronounced overturning at the hub with the second end wall design. So generally the profiling increases the secondary effect at the tip and the second design reduces the hub secondary effect.

The experimentally determined streamwise vorticity is shown in Figure 3.18. At 20% span the positive vorticity peak of the passage vortex is seen **3**, which is decreased with the first design by about 50% compared to the

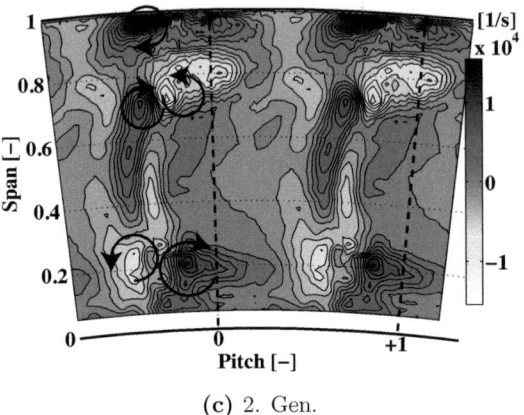

(c) 2. Gen.

Figure 3.18: Measured Streamwise Vorticity ω_s at R1ex in the Rotor Relative Frame

baseline. This results in a reduction of about 10% of the integrated circulation. The integrated circulation is the result of a numerical integration of the vorticity over the area covered by the vortex. However, this doesn't directly translate into a blade row loss reduction at the hub. With the second design the hub passage vortex peak vorticity is also decreased by about 50% compared to the baseline. Additionally the area of the passage vortex with the second end wall geometry is decreased as well, resulting in an integrated circulation reduction of the hub passage vortex by 45% compared to the baseline, which translates into a relative total pressure loss reduction as seen in Figure 3.17 (a). The hub trailing shed vorticity seen as a negative zone **7** to the left of the passage vortex is also reduced with both non-axisymmetric profiles. The other vortical structures are the tip leakage vortex **1** close to the tip and the tip passage vortex of opposite vorticity **2** than the tip leakage vortex. Finally there is the tip trailing shed vortex **8** as region of positive vorticity. Generally the tip flow structure show higher vorticity as a result of positive incidence as described in subsection 3.6.1.

The rotor secondary kinetic energy is evaluated in the relative frame. The secondary kinetic energy at the rotor exit hub is in contrast to the vane exit value the result of the radial secondary flow velocity component and not the circumferential. Therefore, the high kinetic energy is located at the right and left periphery of the vortex as shown in Figure 3.19. The

3.5 Rotor Exit Flowfield

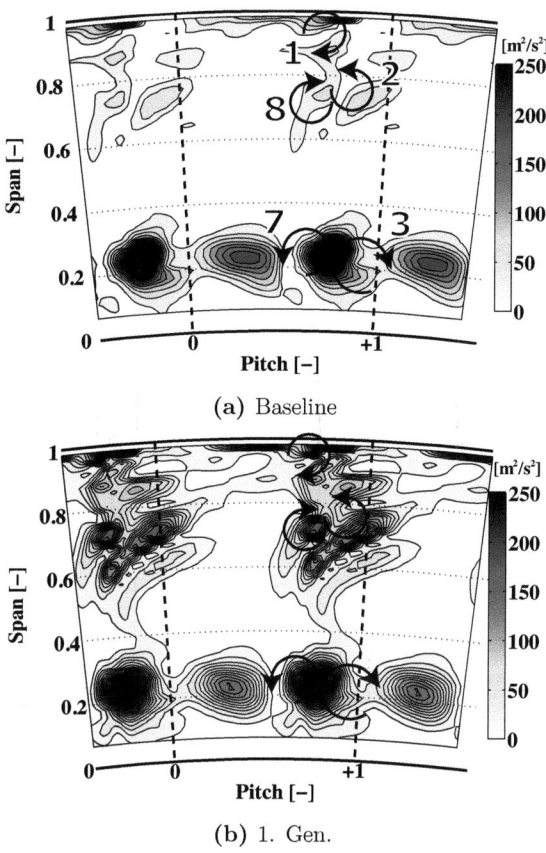

(a) Baseline

(b) 1. Gen.

Figure 3.19: Measured Secondary Kinetic Energy SKE at R1ex in the Rotor Relative Frame

reason for this is that the hub secondary vortices are at about the same radial position and not on top of each other as at the hub of the first vane row exit. The hub region shows an initial reduction with the first end wall profile compared to the baseline. The second end wall geometry reduces the secondary kinetic energy at the hub profoundly. This is consistent with all the results seen so far. In the tip region the first end wall profile shows an increase in secondary kinetic energy. This is the result of higher turning at the blade tip due to positive incidence, which enhances the secondary

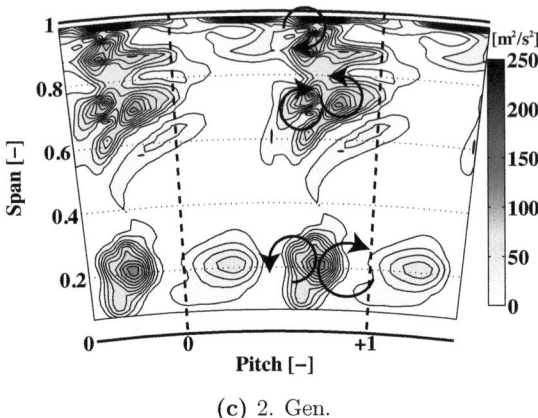

(c) 2. Gen.

Figure 3.19: Measured Secondary Kinetic Energy SKE at R1ex in the Rotor Relative Frame

flows. The second end wall design only shows a small increase in secondary kinetic energy compared to the first design.

3.6 1D Effects of the End Wall Contouring

3.6.1 Change in Reaction

As a result of the profiling the average reaction of the first stage is reduced by 1%. Since the capacity of the first vane didn't change the capacity of the rotor row has to be higher with profiling. The implication is that there is less blockage in the rotor throat. This is both due to suppressed secondary flow at the rotor hub and reduced interaction with the nozzle flow features. It is assumed that this effect is stronger at the hub of the rotor. However, as the rotor is of low aspect ratio and has very high airfoil blockage the hub and tip flowfield is strongly coupled. As a consequence the first vane exit Mach numbers are rising over the whole span with end wall profiling. The absolute Mach number is rising by 1.6% with the first design and by 1.0% with the second design. This will result in more incidence onto the rotor (1°), which will enhance the secondary flow in the rotor row.

3.6.2 Secondary Kinetic Energy & Loss

The prime motivation for non-axisymmetric end wall profiling is a gain in aerodynamic performance of the row profiling is applied. However, due to limitations of practical turbulence models, it is still very difficult to predict the loss during the design phase with CFD. Therefore, as already stated in section 3.1, secondary kinetic energy is used as the prime target function during the design phase. The definition of secondary kinetic energy used in this work is given in section 2.5.3. In this subsection the achieved loss reductions with non-axisymmetric end wall profiling as well as the quality of secondary kinetic energy as a design target quantity are analyzed.

Table 3.3 gives different loss and secondary kinetic energy values. Y S1 Hub is the total pressure loss coefficient which is circumferentially as well as radially mass-averaged from 50% span down to the last measurement point at 4.6% span. Y S1 Tip is the same quantity which is now averaged over a range from 50% span up to the casing. Y S1 Total is the loss value coming from a radial averaging over the whole radial measurement range. Y_{rel} R1 Total is the relative total pressure loss coefficient also radially averaged over the whole radial measurement range. The secondary kinetic energy values (SKE) given in Table 3.3 are non-dimensionalized by the baseline mid-height value. Hub, Tip and Total again refer to the range of the radial mass-averaging.

Variable	Baseline	1. Gen.	2. Gen.
Y S1 Hub [%]	3.27%	3.08%	3.07%
Y S1 Tip [%]	3.87%	3.20%	4.21%
Y S1 Total [%]	3.59%	3.15%	3.67%
Y_{rel} R1 Total [%]	15.1%	14.6%	14.1%
SKE S1 Hub [-]	6.61	4.77	5.03
SKE S1 Tip [-]	8.76	7.78	7.60
SKE S1 Total [-]	7.76	6.46	6.40
SKE R1 Hub [-]	90.9	57.5	23.6
SKE R1 Total [-]	37.6	33.8	20.0

Table 3.3: Measured Loss and Secondary Kinetic Energy Reduction

At the hub both end wall designs result in a loss reduction of about 6% relative to the baseline loss level. At the tip the first end wall design results in a loss reduction of 17% relative to the baseline level while the second design causes a loss increase of 9%. The overall loss level with the first design is 12% lower than with the baseline geometry. The second design results in a loss increase of 2% relative to the baseline case. This discrepancy in the overall first vane loss between the first and second end wall design is only the result of the outer annulus flow field. The overall rotor loss is 3% lower with the first end wall design relative to the baseline case. With the second design the loss reduction is twice as big as with the first design. From these loss numbers it can be concluded that the 1% efficiency increase observed with the first design is predominantly the result of improvements in the first vane. In contrast the efficiency increase of the second end wall design is mainly the result of an improved rotor blade row.

In order to comment on the predictability of secondary kinetic energy the measured reductions of it as given in Table 3.3 are compared to the predicted values achieved during the design calculations. The measured reductions of secondary kinetic energy for the first vane hub end wall relative to the axisymmetric end wall case are 28% and 24% for the first and second design respectively. The prediction shows a close agreement with the predicted values which are 22% for the first design and 19% for the second design. For the rotor hub the trend of reduction is captured by the prediction with 22% for the first design and 33% for the second. However, the measured reductions are more substantial with 37% and 74% for first and second design respectively. At the tip of the first vane neither the trend nor the magnitude of secondary kinetic energy reduction are well capture by the predition. The predicted reductions are 25% and 24% for first and second design respectively. The measured ones are about half as big with 11% and 13% respectively.

If Figures 3.5, 3.6 and 3.7 are compared to the secondary kinetic energy reduction 3.3, it can be seen that there is a good level of agreement between secondary kinetic energy reduction and the reduction of the late passage cross passage pressure gradient. The hub profiling is more successful in reducing the cross pressure gradient than the tip end wall. Therefore, the measured secondary kinetic energy reductions are twice as big at the hub than at the tip. Generally it can be concluded that an effective end wall profile is characterized by a strong cross passage gradient in the early passage and a weakened streamwise perpendicular pressure gradient towards the throat plane.

To assess the quality of secondary kinetic energy as a design target value

3.6 1D Effects of the End Wall Contouring

the relationship between loss and secondary kinetic energy is analyzed. Because in the end it is the loss reduction that counts. Therefore, the measured percentage reductions of loss and secondary kinetic energy are plotted in Figure 3.20.

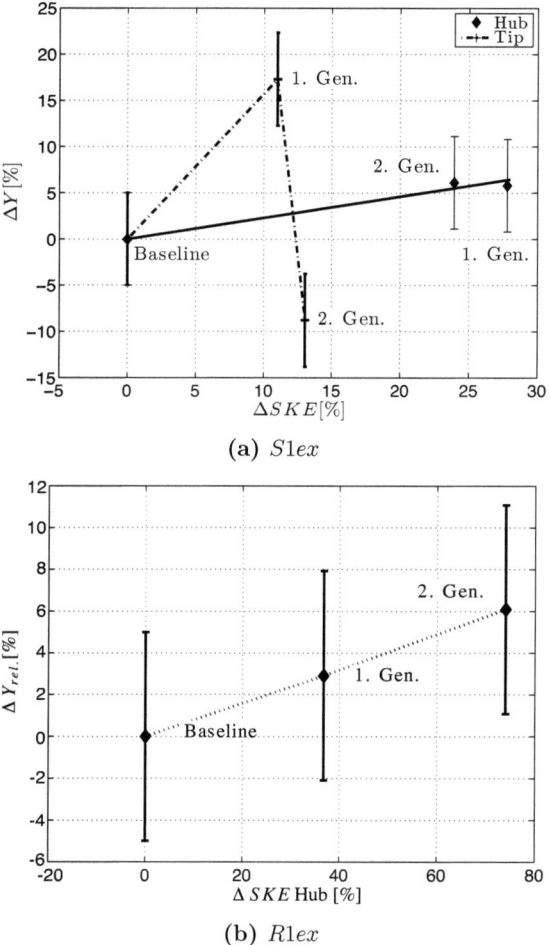

Figure 3.20: Loss Versus Secondary Kinetic Energy for the First Nozzle Guide Vane and the Blade Row

The secondary kinetic energy versus loss reduction for the first vane is plotted in Figure 3.20 (a). At the hub end wall the relationship between loss reduction and kinetic energy reduction is 1% to 5%. For the tip end wall in contrast to the first nozzle hub end wall neither a sensible relationship between secondary kinetic energy and loss can be established nor a reliable prediction of secondary kinetic energy reductions can be made. The reason is that the responsible mechanisms are not well captured by secondary kinetic energy. Despite, the limitations of secondary kinetic energy as optimization parameter it is not recommended to directly use loss as parameter due to the remaining uncertainties concerning turbulence modelling.

Figure 3.20 (b) shows again a linear relationship between loss and secondary kinetic energy reduction for the rotor hub end wall. The ratio between secondary kinetic energy and loss reduction is about 25% to 2%. With the second end wall geometry the relative loss reduction is doubled and is now 6% relative to the baseline. This increase in loss reduction is assumed to be the result of modeling the fillet radii in the design calculations as well as the result of an improved incoming flowfield.

3.7 Time-Resolved Flow Physics

Figure 3.21 shows the downstream absolute frame view at rotor exit of the relative total pressure field of two stator pitches at two different instants in time. One can see the secondary flow features of the three blades as three low relative total pressure zones. From 90% span to the tip one can identify the loss core of the tip leakage vortex labelled as **1** in between 60% and 80% span the loss core of the rotor tip passage vortex **2** and finally between 15% and 35% span the loss core of the rotor hub passage vortex **3**. The two passage vorticies are connected by the rotor wake. The loss cores of the secondary flow features change in shape and intensity depending on their position relative to the upstream and downstream vanes. At $t/T = 0.25$ in Figure 3.21 the labelled secondary loss cores are more elliptical than at the other time instant $t/T = 0$ as they interact with the downstream vane row. Furthermore the wake is strongly bent. The subsequent analysis focuses on three generic regions labelled **A**, **B** & **C**. In order to characterize the three different zones **A**, **B** & **C** two time-space diagrams at mid-height labelled as **Y** in Figure 3.21 are presented in Figures 3.22 and 3.23. It is important to remember that in a stationary frame space-time diagrams the vertically-oriented features are associated with the stator flow field, while rotor flow features show up as inclined structures.

3.7 Time-Resolved Flow Physics

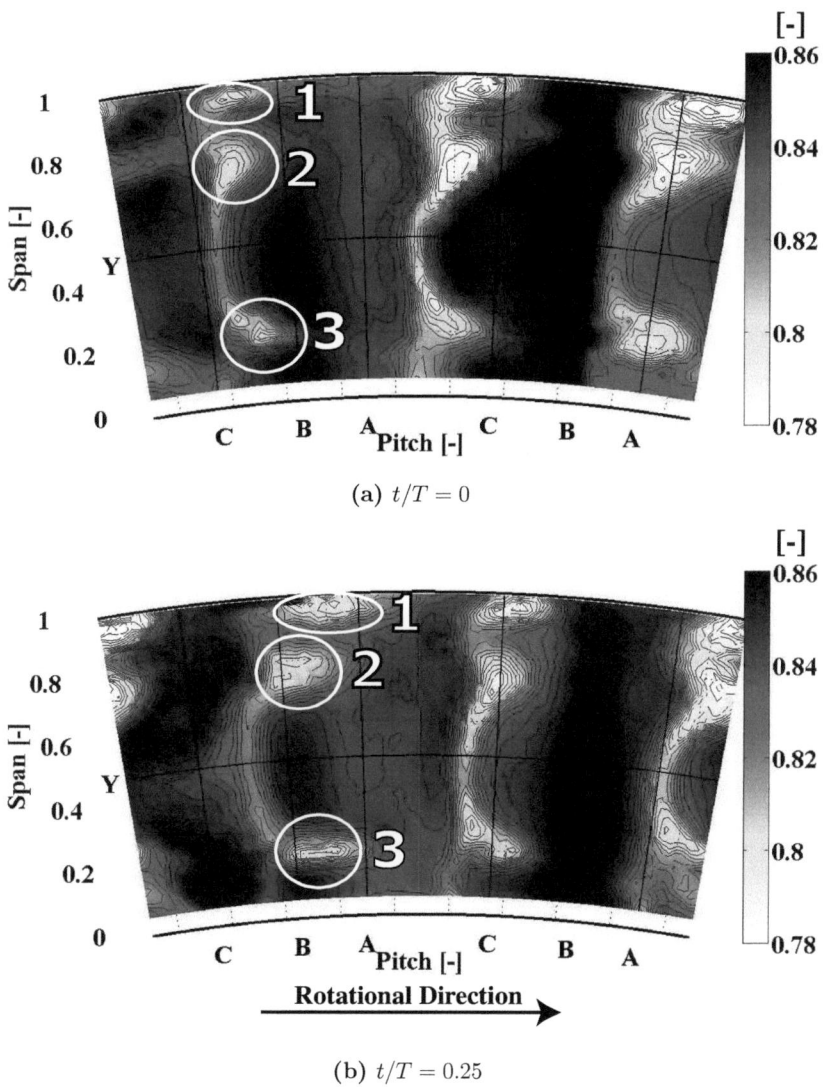

(a) $t/T = 0$

(b) $t/T = 0.25$

Figure 3.21: Measured Relative Total Pressure Coefficient PR_trel at R1ex (Baseline)

Figure 3.22 shows the root mean square values of the random part of the total pressure signal which is a good indicator of flow features. Regions of high **rms** are indicative of eddy shedding or regions of high turbulence. Using the triple decomposition of the time-resolved pressure signal as shown in Equation 3.3 the random part $p'(t)$ can be evaluated as the difference between the raw pressure signal $p(t)$ of the FRAP probe and the phase-locked averaged pressure $\bar{p} + \tilde{p}(t)$. The same approach was used by Porreca [?] to derive turbulent quantities.

$$p(t) = \bar{p} + \tilde{p}(t) + p'(t) \tag{3.3}$$

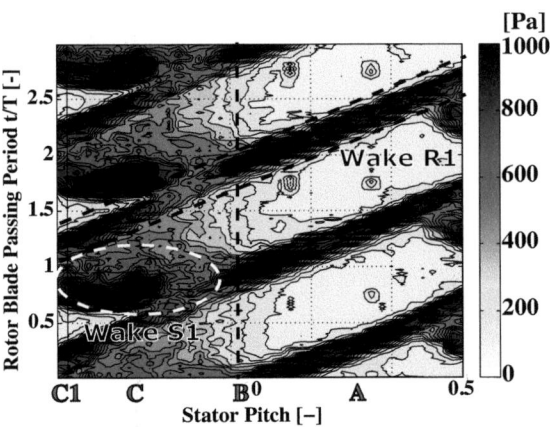

Figure 3.22: Measured Time-Space Diagram at Mid-Height **Y** of **rms** of the Total Pressure Random Part R1ex (Baseline)

In Figure 3.22 three diagonal bands of high **rms** can be seen which are the rotor wakes. The region between these wakes is defined as the free-stream region. Due to the strong migration of the first vane wake fluid from the pressure to the suction side inside the rotor passage as seen in Figure 1.6 only very little wake fluid remains on the pressure side. The result is a zone labelled **A**, which is characterized by low free-stream **rms** values. However, during another interval the vane flow features pass the blade row with minimal blade interaction and end up in zone **C** of Figure 3.22. The result is elevated unsteadiness in the free-stream region. Sharma et al. [?] have introduced the separation of the flowfield into these two zones.

3.7 Time-Resolved Flow Physics

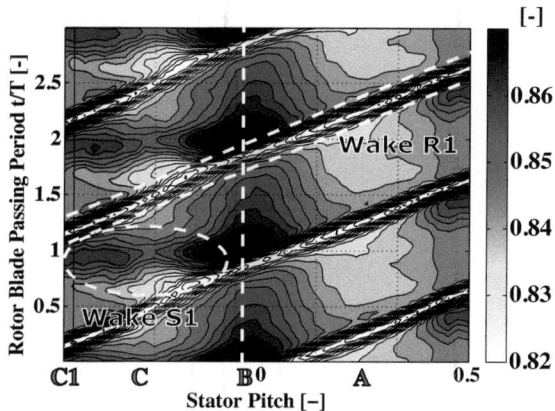

Figure 3.23: Time-Space Diagram at Mid-Height **Y** of the Relative Total Pressure Coefficient PR_trel R1ex (Baseline)

However, if there is a downstream vane row a third zone **B** can be defined which is characterized by high relative total pressure. The second vane row can only have a potential effect. Temporal pressure variations in the relative frame are work processes resulting in a circumferential variation of relative total pressure. Therefore, the region seen in the vicinity of traverse **B** in Figure 3.23 as a dotted line shows elevated relative total pressure. The high **rms** regions around traverse **C** have to be the wake of the first vane for two reasons. Firstly as already stated vertically oriented features are associated with the stationary vanes. Secondly turbulent unsteadiness can only be convected and has therefore, to origin from an upstream component. The region identified as the first vane wake by its **rms** signature is generally of high relative total pressure. If the value at rotor exit is compared to the value a rotor inlet a rise in the relative total pressure of the first vane wake fluid over the rotor is detected. This could have different reasons. Either a radial migration of this fluid or alternatively there is the possibility that in a relative frame consideration work is done on this fluid. This is inline with the ideas presented in Rose et al. [?].

In order to further analyze these three regions, the radial traverses at **A**, **B** and **C** are plotted against time. Figure 3.24 shows the radial-time diagram of the root mean square values **rms** of the random part of the total pressure signal. On the left side the plots for region **A** characterized by the very low free-stream **rms** values. Sharma et al. [?] stated that the rotor secondary flow structures found in this region are very similar to the ones if the blade row would be tested isolated without the influence of upstream or downstream aerodynamic components. The same rotor flow features identified in Figure 3.21 can be seen again as regions of high **rms**. From 90% span to the tip one can identify the tip leakage vortex labelled **1** in between 60% and 80% span the rotor tip passage vortex **2** and finally between 15% and 35% span the rotor hub passage vortex **3**. With the first end wall profile the unsteadiness of the hub vortex is almost unaltered. However, the second end wall profile reduces the unsteadiness of the hub passage vortex considerably (31%). As the rotor is unshrouded there is no end wall profile at the tip of the blade row. Therefore, there are no large differences between the three cases.

Figures 3.25 (a,c,e) show the vorticity of the three main rotor vortices along traverse **A**. **1** the tip leakage vortex rotating anti-clockwise as a result of the pressure to suction over tip leakage. **2** the rotor tip passage vortex rotating clockwise and finally the hub passage vortex **3** which rotates in the same direction as **1**. With the first end wall design the vorticity of the hub passage vortex is reduced. The effect of the second design is a reduction of vorticity as well as a reduction in size.

The right side of Figure 3.24 shows the unsteadiness along traverse **B**. In traverse **B** the unsteadiness in the rotor hub passage vortex rises as the blade vortices are bending around the downstream vane leading edge. However, there are still clear free-stream zones where the **rms** values are very low.

The right side of Figure 3.26 represents the unsteadiness along traverse **C**. Along traverse **C** the integral **rms** values rise once more as additional high **rms** zones occur. As already stated previously these zones have to be the remnants of the upstream vane flow features. There is a high **rms** zone at the hub labelled **4** associated with the upstream vane hub passage vortex. Figures 3.25 (b,d,f) show the streamwise vorticity along traverse **C**. In region **4** a vortex is found which shows the same rotation as the rotor hub passage vortex **3**. Chaluvadi [?] and Behr [?] describe the process of vortex-blade interaction. The process is as follows. The passage vortex of the first vane row is bending around the rotor blade. As a consequence two limbs are evolving. One is traveling along the suction side referred to as the suction side limb. The one which crosses the blade passage is called pressure side limb. The pressure side limb has the same sense of

3.7 Time-Resolved Flow Physics

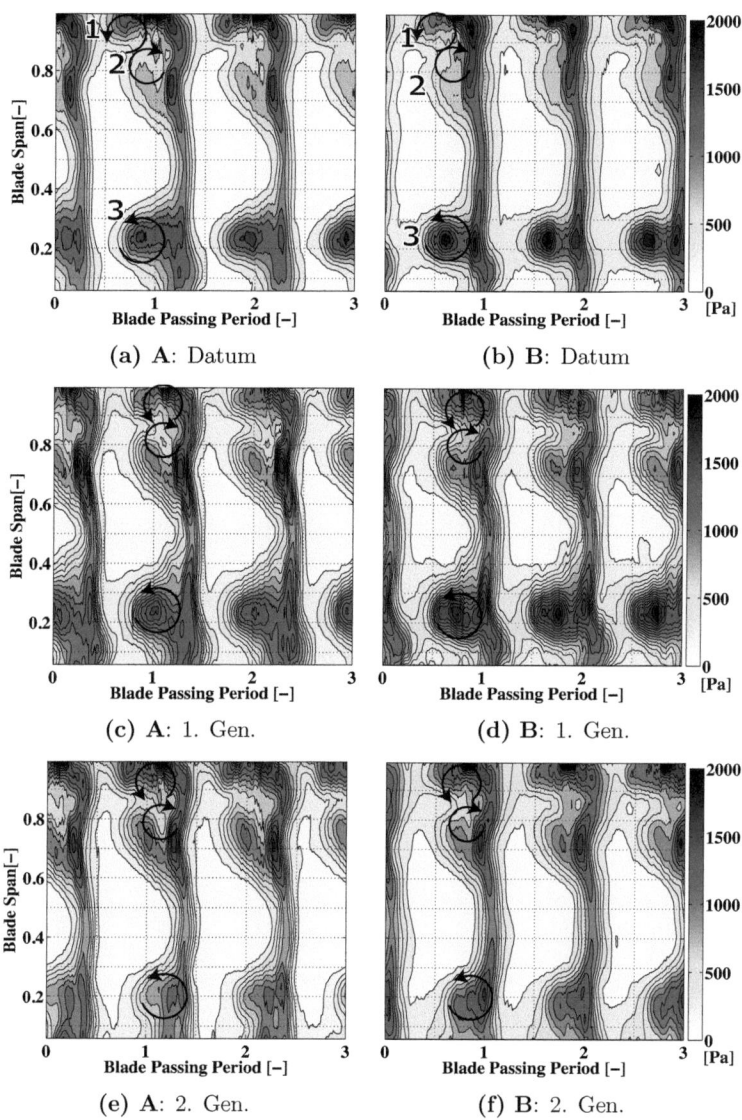

Figure 3.24: Measured **rms** of the Total Pressure Random Part at R1ex

3 Non-Axisymmetric End Wall Profiling

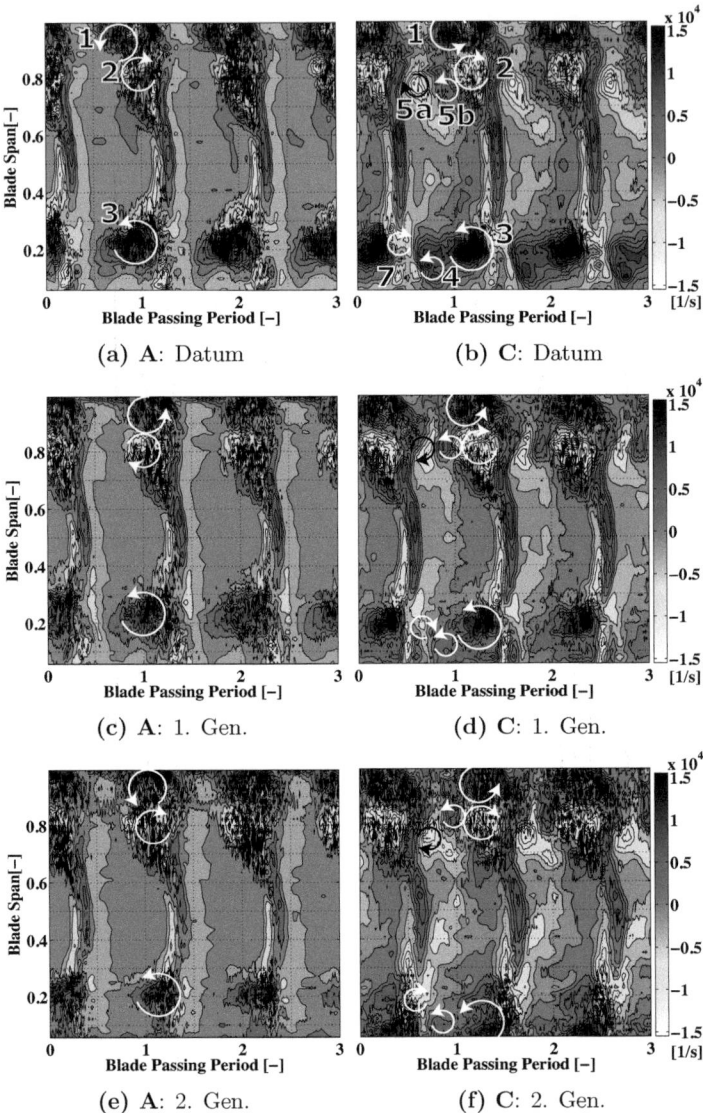

Figure 3.25: Measured Streamwise Vorticity ω_s at R1ex

3.7 Time-Resolved Flow Physics

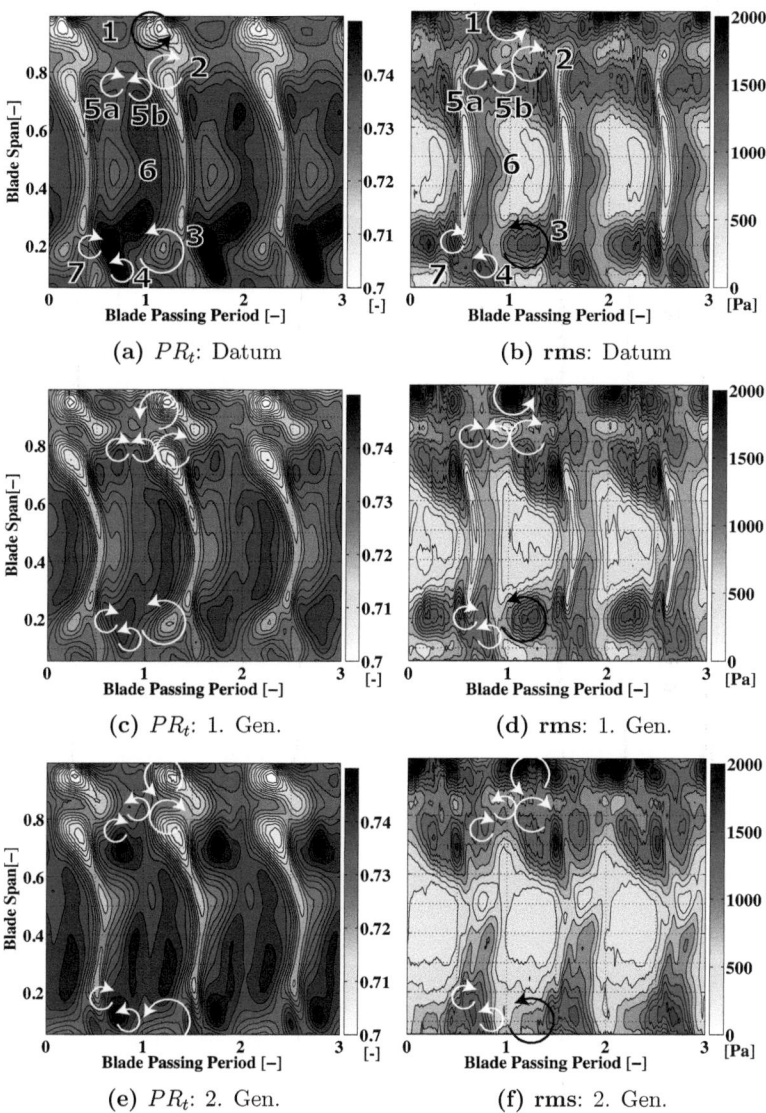

Figure 3.26: Measured Total Pressure Coefficient and **rms** for Traverse **C** at R1ex

rotation as the rotor passage vortex while the suction side limb is rotating in the opposite direction. With this knowledge it can be concluded that vortex **4** is the pressure side limb of the first wake hub passage vortex. The unsteadiness as well as the vorticity of this zone have been decreased with end wall profiling. This is the result of the effective end wall profiling at the hub of the first vane row. At around 75% span there are two counter rotating vortices labelled **5a** and **5b**. Judging from the sense of rotation as seen in Figure 3.25 **5a** is the pressure side limb of the first vane tip passage vortex and **5b** is the suction side limb. In terms of vorticity the first design shows the lowest values for both these vortices. The difference in **rms** is insignificant. The band of high unsteadiness labelled **6** is the first vane wake. Both end wall profiles show reduced **rms** in this region compared to the baseline case. Region **7** is the rotor hub trailing shed vortex.

The left side of Figure 3.26 shows the radial-time diagram of the total pressure for traverse **C**. The rotor loss features are seen as low total pressure zones. However, zone **4** which is seen as region of high **rms** on the right side of Figure 3.26 and identified as the pressure side limb of the first vane hub passage vortex shows a total pressure level higher than the free-stream regions. Also in the vicinity of the first vane tip passage vortex remnants the total pressure level is elevated compared to the free-stream region.

	Datum	1. Gen.	2. Gen.
Freestream	1.0	1.0	1.0
S1 Wake	0.944	0.955	0.944
S1 HPV (PS)	0.950	0.956	0.938
S1 TPV (SS & PS)	0.937	0.945	0.936

Table 3.4: Freestream Relative Total Pressure Ratio μ of First Vane Flow Features

Using the **rms** plots, the first vane and rotor loss cores at the exit of the rotor can be identified. With that the associated rotor exit total pressure values of the first vane flow structures $p_{t,w2}$ can be determined. The freestream total pressure value at rotor exit $p_{t,f2}$ is defined as the total pressure value coincident with the region of very low **rms** in traverse **C**. The rotor inlet freestream total pressure $p_{t,f1}$ is defined as the time-averaged mid-span value at -0.2 pitch. The first vane wake total pressure value at rotor inlet is defined as the minimum value at mid-span $p_{t,w1}$. The up-

3.7 Time-Resolved Flow Physics

stream vane hub and tip passage vortex total pressure values at rotor inlet $p_{t,w1}$ are defined as the minimum values at 10% and 82% span respectively. Using the values $p_{t,w1}$ and $p_{t,w2}$ the total pressure ratio of the first vane flow structures can be approximated. Furthermore the total pressure ratio in the freestream is determined with the values $p_{t,f1}$ and $p_{t,f2}$. Table 3.4 gives the total pressure ratio of the first vane flow structures in relation to the freestream total pressure ratio. The definition of this variable μ is given in Equation 3.4.

$$\mu = \frac{p_{t,w1}/p_{t,w2}}{p_{t,f1}/p_{t,f2}} \tag{3.4}$$

The μ values of the pressure side (**5a**) and suction side (**5b**) limb of the first vane tip passage vortex show no significant difference. Therefore, only the average value is given in Table 3.4. Generally the μ value of the first vane flow structures is between 5% and 6% smaller compared to the freestream. The first vane wake of the first end wall design shows a 1.1% higher μ value compared to the baseline. For the hub passage vortex the μ value with the first end wall design is 0.6% higher compared to the baseline geometry, while it is 0.8% higher for the tip passage vortex fluid. The μ value of the first vane wake and tip passage vortex with the second end wall design is the same as with the baseline geometry. The first vane hub passage vortex shows even a 1.2% lower μ value than the baseline case. This means that there is the highest work extraction out of the first vane flow structures in the first end wall design case. This will contribute to the higher efficiency. These results are consistent with the efficiency numbers given in Table 3.2 and to some extend could explain the much higher efficiency of the first design case. Furthermore, these results show that the upstream nozzle guide vane wake and vorticies generally do less work on the rotor than the freestream, a conclusion which is consistent with the ideas presented in Rose et al. [?]. Finally the more homogenous total pressure flow field will also result in less mixing losses.

So far all flow features of the first vane could be identified apart of the suction limb of the first vane hub passage vortex. In the work of Behr [?] a very nice clocking study is described. A model is proposed, which states that the suction side limb is located underneath the rotor hub passage vortex. This model can be supported by an additional piece of information, which is a plot of the associated vorticity. Therefore, Figure 3.27 shows a radial time-space diagram in the absolute frame location called **C1** in

Behr [?] and also labelled this way in Figures 3.22 and 3.23. In Figure 3.27 the suction side limb is labelled **4b** and seen as structure of opposite vorticity than the rotor hub passage vortex and as a region of elevated unsteadiness. In terms of unsteadiness the first design shows a less intense region compared to the other two cases. In terms of vorticity no large differences can be determined. In terms of the μ value the difference to the pressure side limb is insignificant. The suction side limb of the first vane hub passage vortex is thus shifted by about a quarter vane pitch in the direction of the rotation relative to all the other first vane flow structures, which are concentrated around traverse **C**.

3.8 Summary

The first stage of the 1.5-stage unshrouded axial turbine configuration has been redesigned twice with non-axisymmetric end walls. With the first end wall design the total-to-total efficiency is increased by $1\% \pm 0.32\%$. This increase is mainly the result of an improved first vane flow field. The total pressure loss coefficient is reduced by 6% in the lower annulus half and by 17% in the outer half. In the lower annulus half the loss reduction is the result of reduced secondary flows. However, in the outer half there is also a considerable reduction of wake loss. This is due to an alleviation of vane blockage and consequent reduction of the suction side peak Mach numbers. Furthermore the end wall contour leads to an aft loaded pressure profile which potentially delays transition. The reduction of secondary flow out of the vane means also an improved flow field into the blade row. The first design rotor hub profile has only a small effect. Therefore, the rotor loss coefficient (based on exit relative dynamic head) is reduced by 3% of the relative total pressure loss coefficient relative to the datum geometry ($Y_{rel} = 15.1\%$ in the baseline case).

Using their **rms** signature the first vane flow structures can be identified at rotor exit. Assessing the freestream relative total pressure ratio μ of these structures it can be seen that it is about 5−6% lower than in the freestream. Furthermore the first end wall design shows on average a 1% higher μ value relative to the axisymmetric baseline case. This will contribute to the large improvement in total-to-total efficiency of the first end wall design compared to the baseline case.

With the second design the total-to-total efficiency is increased by $0.3\% \pm 0.32\%$. This improvement is based on a reduction of the first vane row lower

3.8 Summary

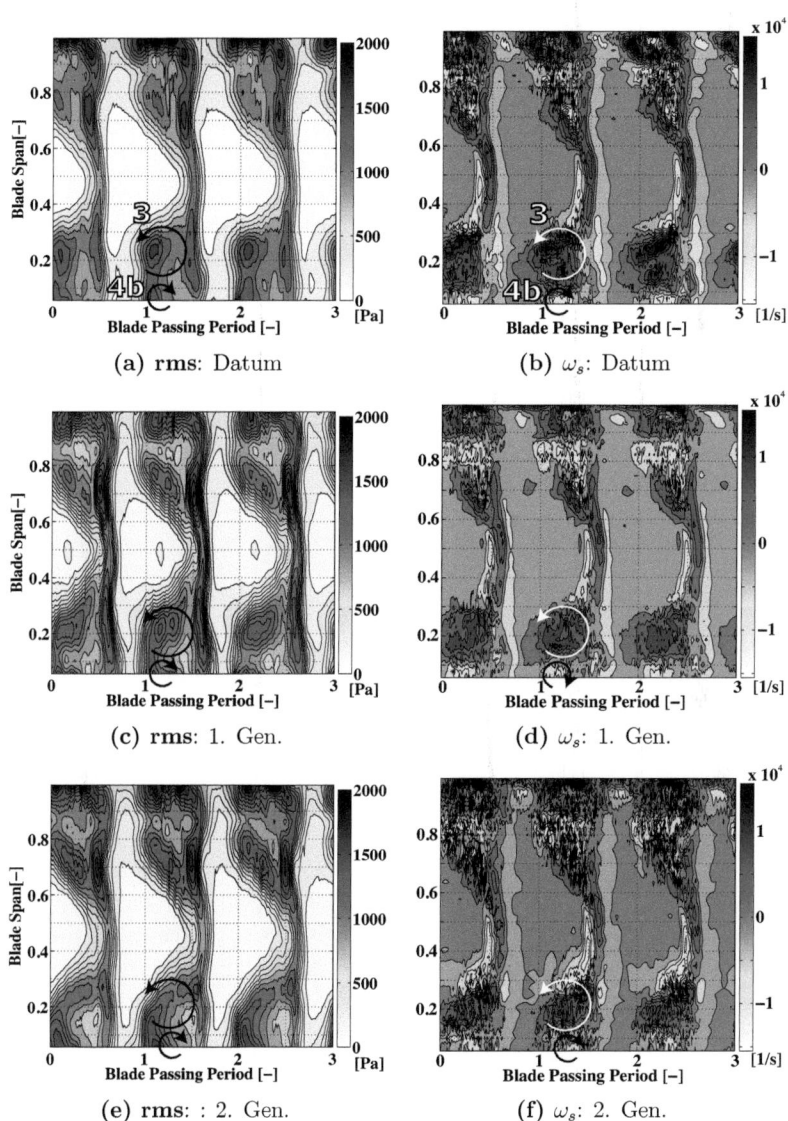

Figure 3.27: Measured Traverse **C1** Suction Side Limb First Vane Hub Passage Vortex at R1ex

half loss coefficient by 6% and a reduction of the rotor loss coefficient by 6%, which is twice as much as with the first design as a result of a strongly reduced hub passage vortex. The further improvement of the rotor end wall is the result of the new design approach which includes fillet radii. This leads to a better prediction of the flowfield with high blockage blading.

The prime reason for the drop in total-to-total efficiency compared to the first design is a much higher mass-averaged loss in the last 15% of the span towards the tip which results in 9% more integrated loss for the outer annulus half relative to the baseline case. It is shown that this is potentially the result of some sort of separation (corner stall) as a result of a to aggressive curvature on the late suction side. Another reason for the lower total-to-total efficiency compared to the first design is a lower total pressure extraction out of the first vane flow structures which is in the range of the baseline geometry.

Due to limitations in turbulence models a direct optimization on loss is not suggested. Therefore, an alternative quantity such as secondary kinetic energy is needed to design end walls. This makes it important to know the relationship between loss and secondary kinetic energy. With the data of the three cases this relationship between secondary kinetic energy reduction and loss reduction can be assessed. The two hub end walls show a linear relation between secondary kinetic energy reduction and loss reduction. At the same time the secondary kinetic energy prediction shows the same trend as the measured values for the two hub end walls. However, the first vane tip end wall shows a low agreement of loss and secondary kinetic energy reduction because the responsible loss mechanisms are not captured by a change in secondary kinetic energy.

Finally the measured first vane hub end wall static pressure fields reveal that an effective non-axisymmetric profile in terms of secondary kinetic energy reduction is characterized by a strong cross pressure gradient in the early passage and a strongly reduced cross passage gradient close to the throat plane. Therefore, it should be tried to add the pressure gradient information as an optimization parameter.

4 Purge Flow Effects

In high-pressure gas turbines, a small amount of purge flow is ejected at the hub rim seal, to cool and prevent the ingestion of hot gases into the cavity between the stator and the rotor disk. In order to comment on the effect of purge flow on the performance of the non-axisymmetric end walls, first a detailed analysis of the secondary flow structures with and without purge flow in the axisymmetric turbine is required. In this chapter the data of two purge flow settings are compared. The baseline case is the suction case with $IR = -0.1\%$. The definition of the injection rate IR is given in Equation 2.1. The standard purge flow setting is an injection rate of $IR = 0.9\%$. The purge flow in this investigation is applied at the rotor upstream rim seal. There is no net massflow over the rotor downstream rim seal. In this chapter the data of the experimental investigation executed in the LISA test rig as well as of a time-resolved CFD calculation are presented.

4.1 Baseline Turbine Purge Sensitivity

In this section the performance behaviour of the axisymmetric baseline turbine is assessed if purge flow at an injection rate of 0.9% of the mainflow is applied instead of small suction ($IR = -0.1\%$).

TR	$T_{t,purge}/T_{t,main}$	0.985
BR	$(\rho \cdot U)_{purge} / (\rho \cdot U)_{main}$	0.110
MFR	$(\rho \cdot U^2)_{purge} / (\rho \cdot U^2)_{main}$	0.013

Table 4.1: Non-Dimensional Injection Parameters at $IR = 0.9\%$

In Table 4.1 the characteristic non-dimensional injection parameters at an injection rate of $IR = 0.9\%$ are given. The parameters are based on measured data.

The definition of efficiency used in this study, accounting for the purge flow is given in Equation 3.2. The resulting turbine efficiency drop is $\Delta \eta_{tt} =$

0.6% ± 0.32% if the injection rate is increased from $IR = -0.1\%$ to $IR = 0.9\%$. This results in a purge flow sensitivity of $-0.6\%/\%$.

4.2 Measured Purge Flow Effect at Rotor Inlet

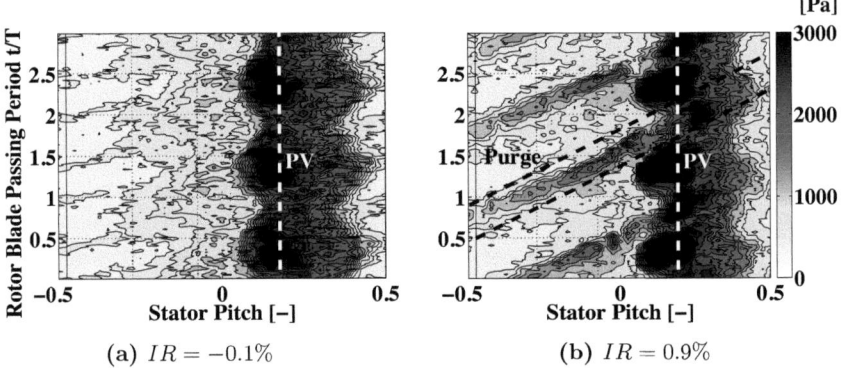

(a) $IR = -0.1\%$ (b) $IR = 0.9\%$

Figure 4.1: Measured Time-Space Diagram at 6% Span of the **rms** of the Total Pressure Random Part (Baseline)

Unfortunately the standard traversing plane at rotor inlet is in the same axial position as the rim seal. Combined with the $1mm$ safety distance of the probes the purge flow is not seen. Therefore, a time-resolved computation is needed to study the purge flow effects and is presented in section 4.4. However, at the lowest radial traverse position some influence of the purge flow is seen. Figure 4.1 shows a absolute frame time-space diagram over one stator pitch and three blade passing periods at 6% span which is the lowest traverse position. The no purge flow case $IR - 0.1\%$ shows a vertical band of high **rms** at 0.2 pitch which is modulated by the rotor. This is the trace of the first vane hub passage vortex. The purge flow case $IR = 0.9\%$ shows three additional inclined bands of high **rms**. As the leakage jets introduce blockage the turbulent end wall fluid is lifting off the end wall. As seen in Figure 4.1 this effect is strongly associated with the rotor.

Figure 4.2 shows the pitch angle in the same representation as the **rms** in Figure 4.1. Positive pitch angle values represent regions of outwards movement and vice-versa. In the suction case $IR - 0.1\%$ there are inclined

4.3 Measured Purge Flow Effect at Rotor Exit

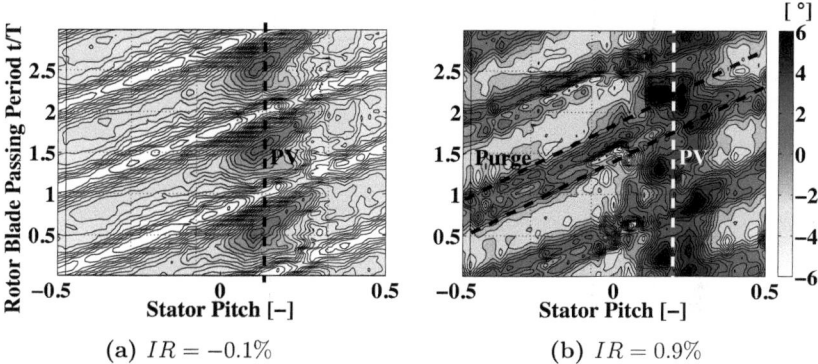

Figure 4.2: Measured Time-Space Diagram at 6% Span of the Pitch Angle γ (Baseline)

bands of negative pitch angle. Most probably where the rim seal is sucking. In the region of the hub passage vortex the flow is pitching out. With purge flow $IR = 0.9\%$ the trend is reversed. In the region of the high **rms** bands the fluid is pitching out as there are positive values. Overall it can be assumed from the measurements that the injection jets are most probably traveling at blade speed. At least their upstream effect suggests it.

4.3 Measured Purge Flow Effect at Rotor Exit

4.3.1 Time-Averaged

Figure 4.3 and 4.4 show the influence of the purge flow at rotor exit in terms of radial distribution of deviation and efficiency. Figure 4.3 shows the circumferentially massflow averaged deviation from the metal angle angle at the exit of the rotor. The design metal angle is $-67°$ and constant over the whole span. At the hub one can see the underturning region between 20% and 30% span. At 10% span there is the overturning region induced by the hub passage vortex. At 72% span the tip passage vortex introduces underturning. At 88% span there is the combined overturning of the tip passage vortex and the tip leakage vortex. Finally at the casing the underturning part of the tip leakage vortex can be seen. In the outer half of the annulus the influence of the injection is small. At the hub

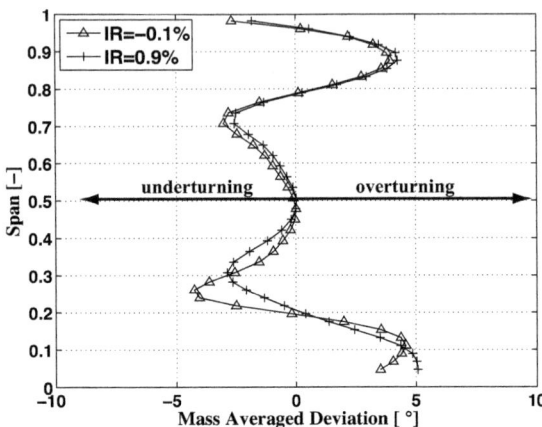

Figure 4.3: Measured Mass-Averaged Relative Flow Angle Deviation Relative to the Mid-Height Value at R1ex

the underturning is reduced by 1.4° with purge flow, while it has radially migrated outwards by 5% of the span. The overturning is the same in both cases. With purge flow the highest overturning is at a 5% lower span position. With purge flow the signature of the secondary flow is generally diminished, while it has a larger radial extent.

Figure 4.4 shows the radial distribution of the total-to-total efficiency. It should be noted that this efficiency is linearly related to the total pressure distribution as the power is a $1D$ value calculated on the basis of a torque tube measurement. With purge flow there is an increasing efficiency deficit towards the hub. The suction case has a pronounced efficiency deficit peak around 20% span. In the purge flow case the efficiency deficit is distributed over a larger radial extend.

Figure 4.5 shows the relative total pressure at the rotor exit time-averaged in the rotating frame of reference. Below 30% span there is the low relative total pressure zone that is associated with the rotor hub secondary loss **3**. Between 65% and 80% the tip secondary loss core can be identified **2**. Between 90% span and the casing there is the signature of the tip leakage vortex **1**. The narrow band of low relative total pressure between 30% and 65% represents the rotor wake. In the purge flow case the hub loss zone **3** is about twice the size of the zone in the sucking case. This loss zone extends from 40% span down to the measurement line closest to the hub.

4.3 Measured Purge Flow Effect at Rotor Exit

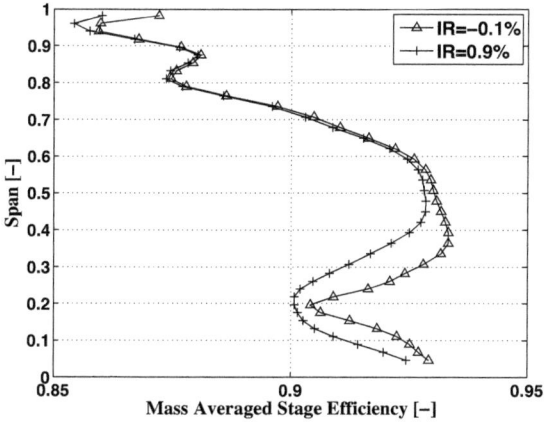

Figure 4.4: Measured Mass-Averaged Total-to-Total-Efficiency η_{tt} at R1ex (Baseline)

However, the loss peak is reduced.

Equation (4.1) gives the definition for the relative total pressure loss coefficient Y_{rel}. In Figure 4.6 the radial distribution of Y_{rel} is given. Y_{rel} is calculated using the mass-averaged rotor inlet and exit relative total pressure values at the same radial height. Figure 4.6 confirms the observations made in Figure 4.5. The peak loss in both cases is the same to within one percent. The purge flow however, shows loss over a much larger radial extent, resulting in a 18% higher integrated loss in the lower annulus half.

$$Y_{rel}(r) = \frac{p_{trel,S1ex}(r) - p_{trel,R1ex}(r)}{p_{trel,R1ex}(r) - p_{s,R1ex}(r)} \cdot 100 \qquad (4.1)$$

In Figure 4.7 the streamwise vorticity at the rotor exit time-averaged is shown in the rotating frame of reference. The positive streamwise vorticity seen at 20% span and labelled **3** is due to the rotor hub passage vortex. The negative vorticity in region **7** can be associated with the hub trailing shed vortex. At 75% the vorticity of the tip passage vortex can be seen in region **2**. The tip trailing shed vortex is labelled **8**. Along the casing the positive vorticity signature of the tip leakage vortex is seen in region **1**. With purge flow the peak vorticity of the hub passage vortex is about halved. However, the integrated circulation is 10% higher with injection. The integrated circulation of the hub trailing shed vortex has increased by

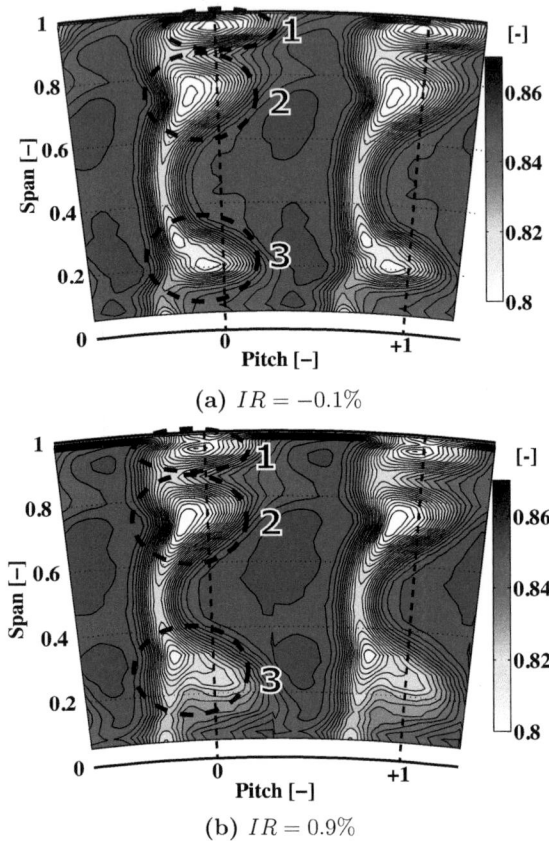

Figure 4.5: Measured Time-Averaged Relative Total Pressure Coefficient PR_trel in the Rotor Frame of Reference at R1ex (Baseline)

30%. This indicates a greater variation in blade lift with span. Therefore, the lift at the hub must have reduced. The purge flow shows no significant influence in the outer annulus half.

Figure 4.8 shows the time-averaged **rms** values of the random part of the total pressure signal in the rotating frame of reference. There are distinct **rms** peaks associated with the secondary vortices. The hub passage vortex **3** as well as hub trailing shed vortex (below 40% Span), the tip passage vortex **2** as well as the tip trailing shed vortex **8** (between 60% and 80%

4.3 Measured Purge Flow Effect at Rotor Exit

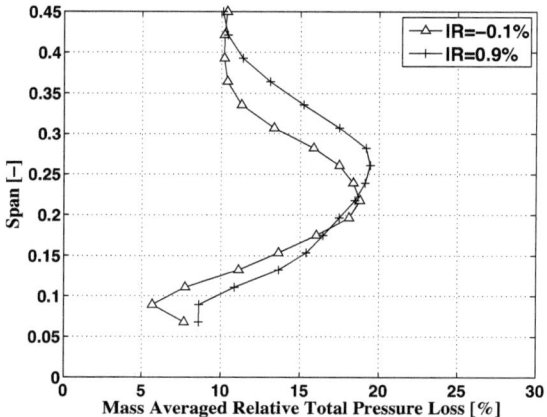

Figure 4.6: Measured Time-Averaged and Circumferentially Mass-Averaged in the Rotor Frame of Reference Relative Total Pressure Loss Y_{rel} at R1ex (Baseline)

Span) and the tip leakage vortex **1** (beyond 90% span). Furthermore there is a band of high **rms** values between 40% and 60% span associated with the rotor wake. The **rms** plot shows a very diffused hub passage vortex region with purge flow, which has increased in size and reduced in peak intensity compared to the sucking case. The integrated hub **rms** value has increased by 30% with purge flow.

4.3.2 Time-Resolved

Figure 4.9 shows a downstream absolute frame view of the relative total pressure field for the blowing case $IR = 0.9\%$ for two stator pitches at exit of the rotor. As the vane to blade ratio is two to three, the loss systems of three rotor blades can be seen. Figure 4.9 (b) shows the situation at same instant in time as Figure 3.21 but now for the purge flow case. The rotor loss features can be again identified as the low relative total pressure zones. From 90% span up to the casing the loss is associated with the tip leakage vortex labelled with **1**. Between 65% and 80% span the tip passage vortex loss **2**. In between 30% and 65% span the rotor wake can be identified. From 10% to 30% span the low relative total pressure zone shows the rotor hub passage vortex **3**. The low momentum purge flow is entrained by the

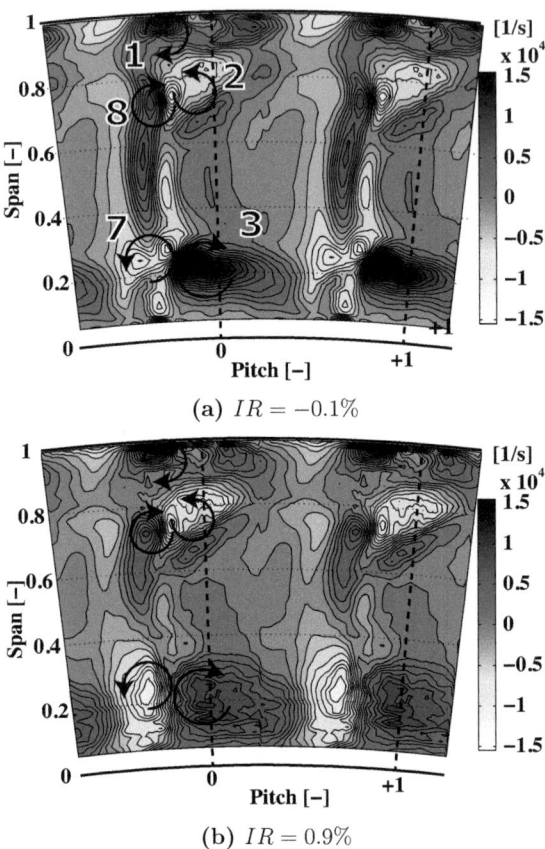

Figure 4.7: Measured Time-Averaged Streamwise Vorticity ω_s in the Rotor Frame of Reference at R1ex (Baseline)

hub passage vortex of the rotor. This results in a large and diffused region at the hub. Furthermore the hub passage vortex is less stable than without purge flow. In region **B** the hub passage vortex is almost disappearing.

As shown in section 3.7 the time-resolved rotor exit flow field can be divided into three different zones. The centerlines of these regions are seen as radial lines in Figure 4.9 labelled with **A**, **B** and **C**. Region **A** is the zone characterized by a low free-stream turbulence or unsteadiness. The remnants of the first vane flow structures are concentrated around traverse

4.3 Measured Purge Flow Effect at Rotor Exit

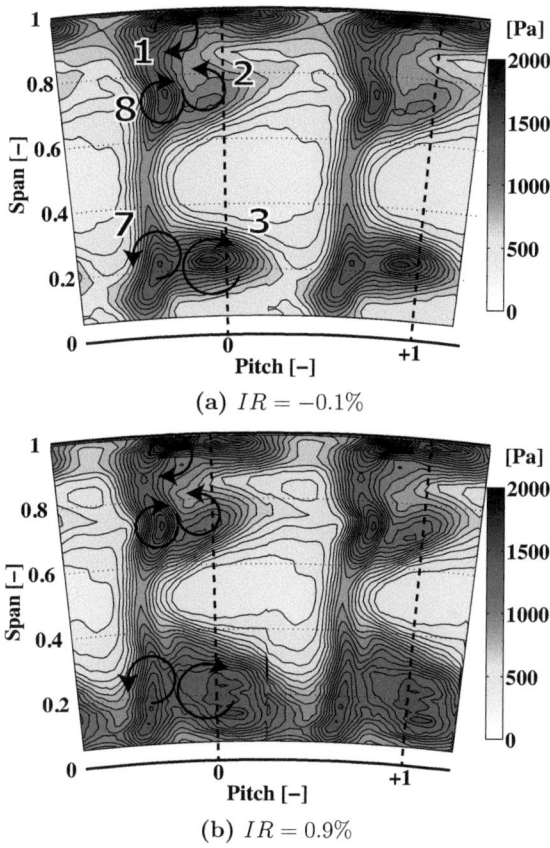

Figure 4.8: Measured Time-Averaged **rms** of Total Pressure Random Part in the Rotor Frame of Reference at R1ex (Baseline)

C. In traverse **B** the strongest influence of the downstream vane row is seen. As already stated in section 3.7 the downstream vane row has only a potential effect. The bow waves of the downstream vane leading edge are seen as a region of elevated relative total pressure in Figure 4.9.

In order to study the unsteady effects of the purge flow the three traverses **A**, **B** and **C** are plotted against time. Figure 4.10 shows the **rms** values of the random part of the total pressure signal as a plot of radius against time for the three traverses. Figure 4.10 (a) and (b) show the lowest influence of

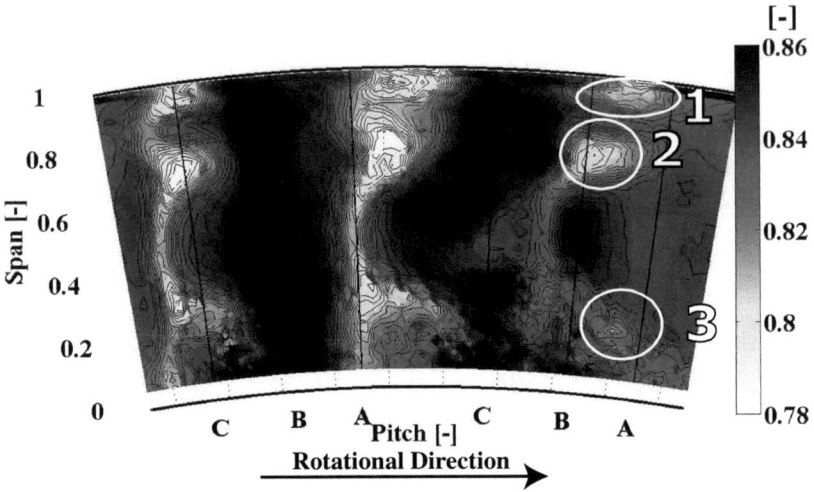

Figure 4.9: Relative Total Pressure Coefficient PR_trel for $IR = 0.9\%$ at R1ex at $t/T = 0$ (Baseline)

the vane rows on the rotor flow features characterized by the lowest integral **rms** values. The rotor flow features as identified in Figure 4.9 can be seen as high **rms** regions. From 90% span to the tip the tip leakage vortex unsteadiness labelled **1** is found. In between 60% and 80% span there is the unsteadiness of the rotor tip passage vortex **2**. The unsteadiness of the rotor hub passage vortex **3** is located between 15% and 35% span. With purge flow the high unsteadiness of the hub passage vortex is closer to the hub and spread over a larger area compared to the suction case.

In traverse **B** the unsteadiness in the rotor hub passage vortex rises due to the interaction with the downstream stator leading edge. The purge flow case shows a stronger increase in **rms** compared to traverse **A**.

Along traverse **C** the integral **rms** values rise once more as additional high **rms** zones are occurring. These zones show the remnants of the upstream vane flow features as discussed in section 3.7. There is the high **rms** zone at the hub labelled **4** associated with the upstream vane hub passage vortex. With purge flow this zone is much less pronounced than in the suction case. The high **rms** zone of the rotor hub passage vortex **3** is lifting off the end wall once it starts to interact with the upstream vane hub passage vortex **4**. This is probably happening as the leakage fluid has very little momentum.

4.3 Measured Purge Flow Effect at Rotor Exit

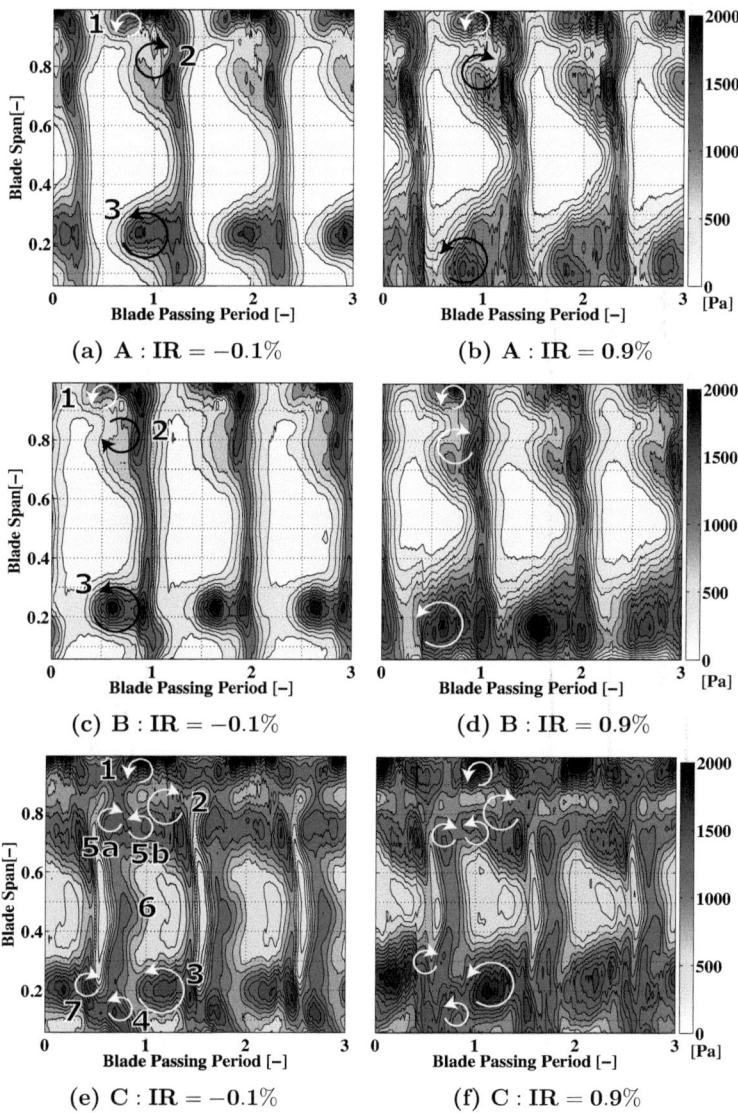

Figure 4.10: Measured **rms** of Random Part of Total Pressure at R1ex (Baseline)

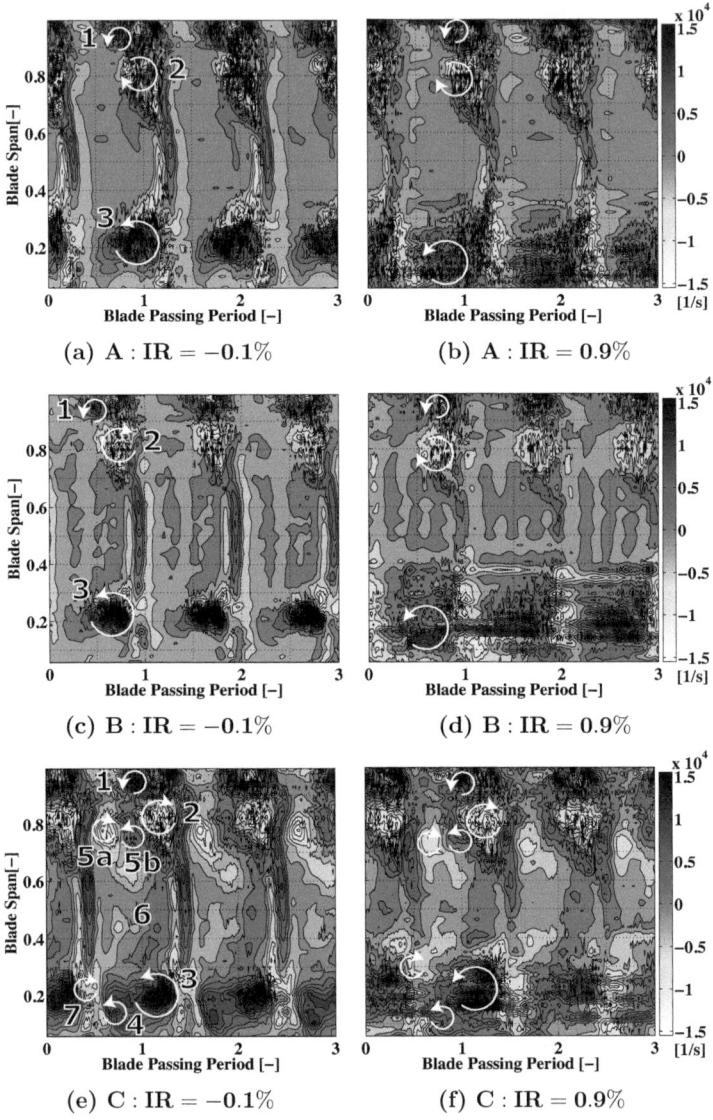

Figure 4.11: Measured Streamwise Vorticity ω_s at R1ex (Baseline)

4.3 Measured Purge Flow Effect at Rotor Exit

Figure 4.11 shows the streamwise vorticity as a plot of radius against time for the same three traverses at rotor exit. In Figure 4.11 (a) and (b) the vorticity of the three rotor secondary flow structures are seen. **1** the tip leakage vortex, **2** the tip passage vortex and **3** the hub passage vortex. Independent of the the vane relative position the two tip vortices are hardly influenced by the injection. With purge flow the hub passage vortex is closer to the hub by about 6% of the span and the radial extend of the hub passage vortex is twice as large compared to the suction case. At the same time the vorticity is diminished with purge flow.

In traverse **B** the rotor vortices are interacting with the downstream vane leading edge. In the suction case this means that the vortex is bent around the leading edge of the downstream vane. As a result the hub passage vortex tube becomes smaller. In order to maintain the circulation the vorticity has to increase as seen in Figure 4.11 (c). In the purge flow case the behaviour is different. The hub passage seems to cover a larger area and to be weaker in terms of vorticity. Combined with the observed strong increase in **rms** at traverse **B** (Figure 4.10 (c)) it can be argued that the rotor hub passage vortex is probably not bent but rather chopped by the vane and as a result breaking down. Furthermore with purge flow the hub passage vortex center is migrating radially out by 2% of the span relative to traverse **A**. In the suction case the center of the vortex remains at the same radial position.

As seen along traverse **C** the tip flow features of the first vane are not really influenced by the purge flow. However, as already seen in the **rms** plots there is no clear first hub passage vortex flow structure **4** in the purge flow case. Probably it has as well merged with the rotor hub passage vortex. With purge flow the rotor hub passage vortex is again moving radially out by 5% of the span relative to position **B**. In the suction case the center of the vortex still remains at the same radial position. At this circumferential position the vorticity of the rotor hub passage vortex is the highest with purge flow.

4.3.3 Spectral Analysis

The frequency composition of the flow in the hub loss core is determined from an FFT of the raw voltage signal of sensor 1 of the FRAP probe, which is directly related to the pressure signal. The amplitude is non-dimensionalized by the blade passing value at mid-height of the suction case. The measurements at 16% span and -12.5% pitch are considered.

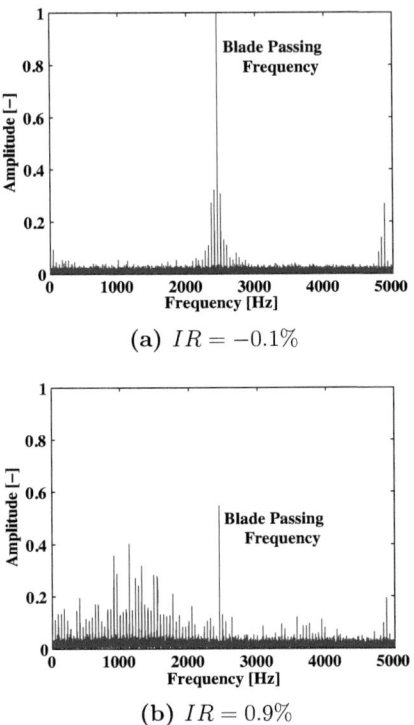

Figure 4.12: Measured FFT of the Raw Voltage Signal of the FRAP Probe at 16% Span and -12.5% Pitch (Baseline)

The results are presented in Figure 4.12. The blade passing period is $2430Hz$. With $IR = -0.1\%$ there is a distinct peak at the blade passing frequency. With purge flow the peak amplitude at the blade passing frequency is reduced. Furthermore there is a band of relatively high amplitudes in the subharmonic spectrum of the blade passing frequency. As observed by Boudet et al. [?] purge flow can introduce such subharmonic frequencies. They found that these frequencies result from a non-linear interactions with cavity instabilities. In the present work no time-resolved cavity data is available. Therefore, no conclusive statement on the origin of these frequencies can be made. Alternatively there are possibly some non-deterministic modes of the rotor hub secondary flows influenced by the

4.4 Time-Resolved Calculation

purge flow. The subharmonic frequencies show an effect in the streamwise vorticity time-space diagrams as seen in Figure 4.11. The purge flow case shows a banding in radial direction. This is an artifact of the phase-lock averaging if subharmonic frequencies are present. As the result of the subharmonic variation the same phase relative measurement doesn't necessarily show the vortex to be in the same radial location. As a result the phase-lock averaged data does show three bands which are representing regions of an elevated probability for the vortex to be and not a coherent vortical structure as seen in the suction case.

4.4 Time-Resolved Calculation

Since measurements inside the blade row as well as a measurement of the rim seal exit pressure field are not possible the missing information has to be acquired using a computational model. The details of the model as well as the validation can be found in section 2.7.

4.4.1 Static Pressure at the Rim Seal Exit

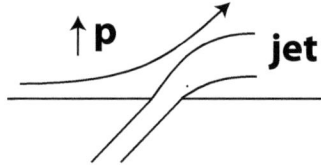

Figure 4.13: Illustration of Blockage Effect of the Injection Jet

The pressure p at the rim seal exit varies much more than the driving pressure inside the cavity $p_{t,cav}$. Therefore, the rim seal exit pressure to a first order is the driving force of the rim seal inflow and outflow. If the pressure is lower than inside the cavity the seal blows and vice versa. The pressure field is an unsteady superposition of the vane and blade pressure fields. However, the pressure field is also influenced by the injection itself as there is a feedback mechanism. When a jet leaves the rim seal it introduces blockage. Therefore, the pressure in front of the jet is increasing as a result of streamline curvature. Figure 4.13 shows an illustration of the effect of additional blockage caused by the injection jet. The rim seal exit pressure is non-dimensionalized by the main flow dynamic head at mid-height. The definition of the resulting non-dimensional pressure quantity ζ is given in Equation 4.2.

$$\zeta = \frac{\overline{p}_{t,cav} - p}{\overline{p}_{t,main} - \overline{p}_{s,main}} \qquad (4.2)$$

In order to visualize the resulting in- and outflow at the rim seal exit the non-dimensional flow quantity ζ is axially averaged over the rim seal and is plotted in an absolute frame time-space diagram as seen in Figure 4.14. The maximum pressure variation in axial direction is about $4\% \pm 0.2\%$. Two vane pitches are plotted against one blade passing event. The vane to blade ratio is two to three.

Values above zero are indicative of outflow and vice-versa. If Figure 4.14 (a) is consulted one sees that the in- and outflow based on the static pressure is characterized by four vertical high and low pressure bands. In absolute frame time-space diagrams vertical structures are associated with the vane rows. Therefore, it can be concluded that the vane pressure field is dominant if no purge flow is applied. However, there is some influence of the rotor. Therefore, especially the high pressure region (darker) is modulated in vertical direction. If purge flow is present the inclined structures become more pronounced. Inclined structures in an absolute frame time-space diagrams are always associated with the blade row. Therefore, the rim seal pressure field with purge is dominated by the rotor. This is most probably the result of the feedback mechanism described further up.

4.4.2 Purge Flow Mechanism

In order to visualize the purge flow an isosurface of the rotary stagnation temperature at $319K$ is shown in Figure 4.15. Since the injected fluid has a $5K$ lower stagnation temperature the temperature isosurface is a valuable visualization of the injection jet shape. In order to provide better conservation with the influence of radial migration of the injected fluid rather the rotary stagnation than the relative stagnation temperature is chosen and given in Equation 4.3.

$$T_{trot} = T_t - \frac{U \cdot C_\theta}{c_p} \qquad (4.3)$$

Figure 4.15 shows two different time steps. As seen in Figure 4.14, the

4.4 Time-Resolved Calculation

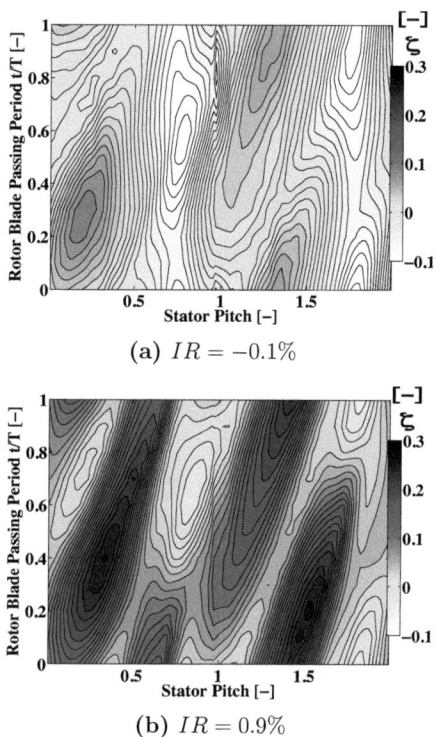

(a) $IR = -0.1\%$

(b) $IR = 0.9\%$

Figure 4.14: Predicted Time-Space Diagram of Axially Average Rim Seal Exit Non-Dimesional Static Pressure (Baseline)

rim seal exit static pressure is dominated by the rotor. Therefore, one can expect that the injected fluid travels at rotor speed or vice-versa is stationary in the rotor frame. This is confirmed by Figure 4.15 as both time steps show the jets to be in the same blade relative position which is in the suction side corner pointing into the blade passage. However, the injection jets are modulated by the stationary pressure field as seen in Figure 4.15. Therefore, the shape and size of the injection jets change.

Figure 4.16 (b) shows a meridional plane in the baseline case which is cutting the injection jet. Figure 4.16 (a) shows for comparison the same plane at the same instant in time for the case without injection $IR = -0.1\%$. In Figure 4.16 (a) there is a distinct zone of positive circumferential vortic-

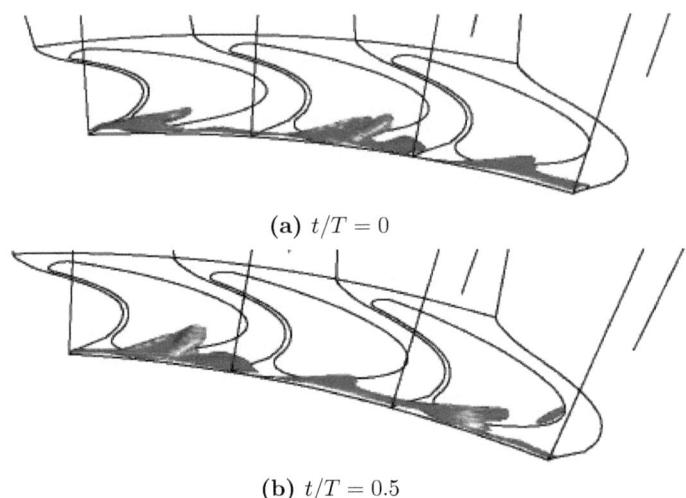

Figure 4.15: Computational Isosurface of Rotary Stagnation Temperature $319K$ with $IR = 0.9\%$ (Baseline)

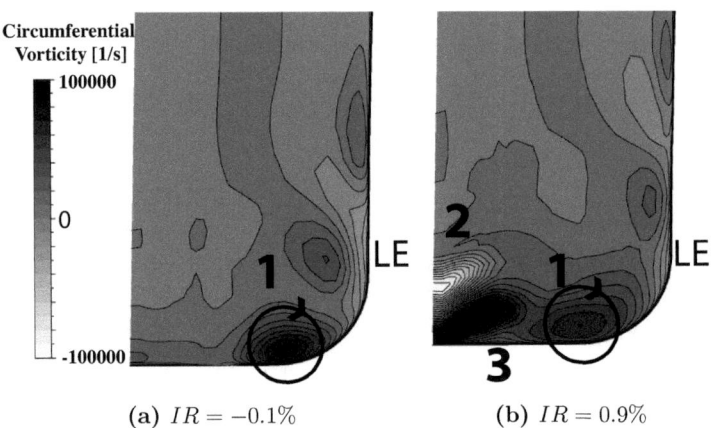

Figure 4.16: Computational Circumferential Vorticity in a Meridional Plane Cutting Through the Injection Jet (Baseline)

ity, labeled **1**, which is the horseshoe vortex. In Figure 4.16 (b) there are additional zones of positive and negative circumferential vorticity resulting

4.4 Time-Resolved Calculation

from the injection. Therefore, it can be concluded that the injection jets are causing zones of normal vorticity of opposite sign as they leave the rim seal. When this additional vorticity created by the purge flow is turned around the rotor blade leading edge a streamwise vorticity component is created.

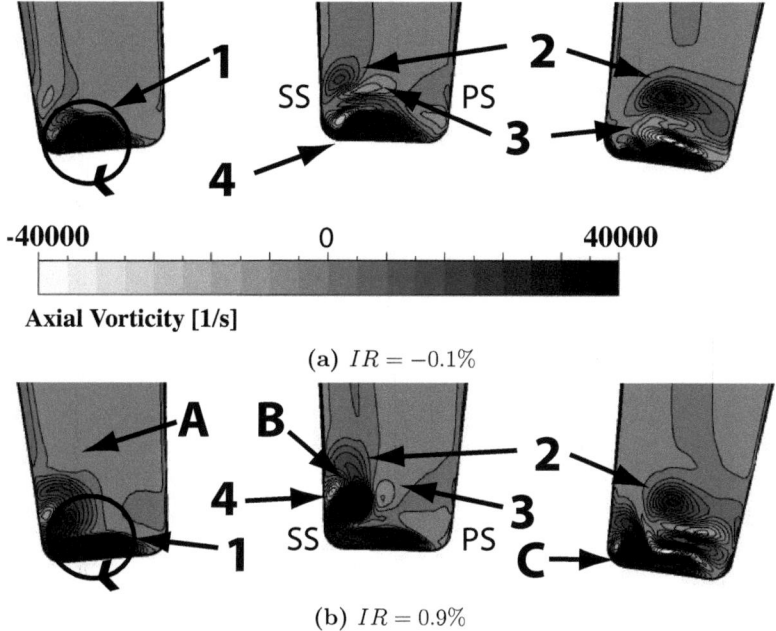

(a) $IR = -0.1\%$

(b) $IR = 0.9\%$

Figure 4.17: Computational Axial Vorticity in an Circumferential Plane at 35% Rotor Axial Chord $t/T = 0$ (Baseline)

In order to assess the streamwise vorticity strength the axial vorticity is plotted in Figure 4.17 in a circumferential plane at 35% rotor axial chord downstream of the leading edge. In this plane the flow is nearly axial therefore, it is adequate to plot the axial vorticity component. Figure 4.17 shows up to five vortical regions depending on the injection mode and the vane relative position of the rotor passage. The vortex labelled **1**, is the pressure side limb of the rotor horseshoe vortex. When the hub passage vortex of the first vane row is wrapping around the rotor leading edge a pressure and suction side leg are evolving seen as regions **2** and **3**. The pressure side leg **2** has the same sense of rotation as the rotor horseshoe vortex pressure side leg. On the suction side there is a region of negative axial vorticity which is the suction side leg of the rotor horseshoe vortex **4**. With injection a very pronounced zone of positive vorticity, having the same sense of rotation as the rotor passage vortex, appears in all three passages simultaneously at the blade suction side above the rotor horseshoe vortex pressure side leg labelled **A**, **B**, **C**. At $t/T = 0$ the structure is most pronounced in the central passage **B**, which corresponds to strong blowing as seen in Figure 4.15 (a). In this situation the suction side leg of the rotor horseshoe vortex **4** is pushed high up the suction side.

As reported by Dubief and Delcayre [?] Q-isosurfaces turn out to display well coherent vortex structures. Q is defined as:

$$Q = \frac{1}{4}\left(\omega^2 - 2S^2\right) \qquad (4.4)$$

ω is the vorticity vector and S^2 the strain rate. The value of Q has to be positive according to the Q criterion, which is a necessary condition for the existence of low pressure vortical tubes. Figure 4.18 shows isosurfaces of $Q = 10^7 [1/s^2]$ of three blade passages in a downstream view with contours of the circumferential vorticity. Additionally, the axial plane at 35% axial chord is plotted. In Figure 4.18 (a) the rotor blade horseshoe vortex **1** can be seen for the suction case $IR = -0.1\%$. Region **2** shows the horseshoe vortex of the following blade. Figure 4.18 (b) shows the purge flow case $IR = 0.9\%$. The vortices created by the injection can be seen in regions **3** and **4** of Figure 4.18 (b), which are absent in Figure 4.18 (a). **4** is the vortex rotating anticlockwise in the meridional plane as seen in Figure 4.16. Structure **3** shows clockwise rotation which is the same sense of rotation as the rotor horseshoe vortex. In the 35% axial chord plane the influence of the anticlockwise vortex , **4** is much stronger than the influence of vortex **3** as seen in Figures 4.17 and 4.18 (b).

As seen on the right side of Figure 4.19 the vortex generated by the purge flow is merging with the rotor hub passage vortex. Figure 4.19 shows the absolute vorticity for three different circumferential planes at 35%, 59% and 83% axial chord (C_{ax}). The suction case on the left side is compared to the purge flow case on the right side. At 35% axial chord the pressure side limb of the horseshoe vortex (PS-HV) is seen as it migrates from the pressure to the suction side without the influence of purge flow. However, in the purge flow case there is another vortex further up on the blade suction side which originates from the injection as described further up (LV). As

4.4 Time-Resolved Calculation

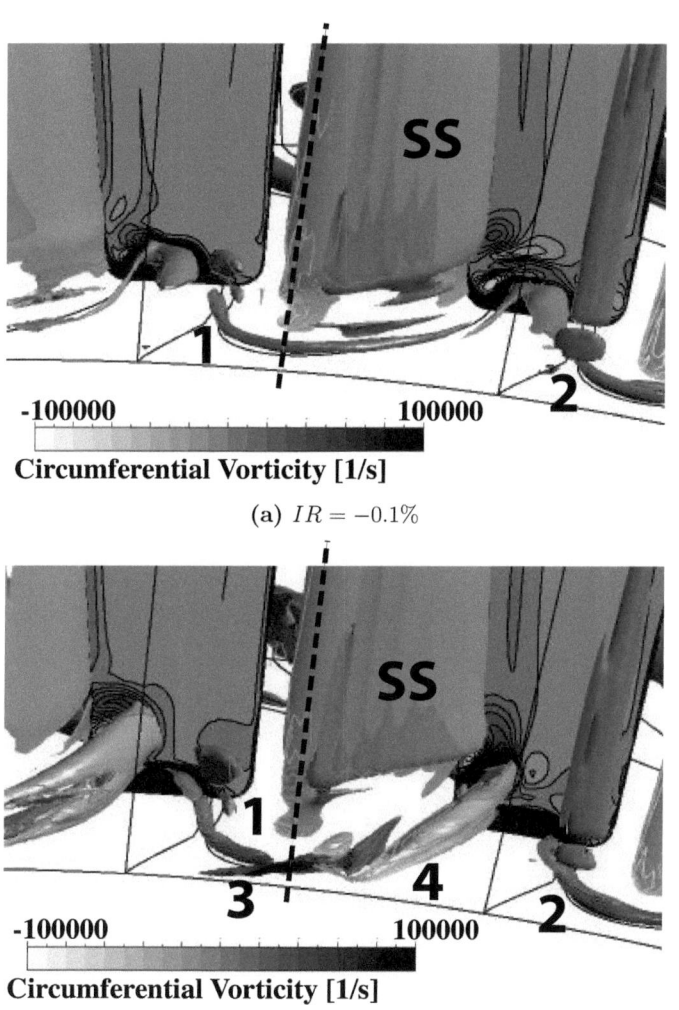

Figure 4.18: Computational Q-Factor $10^7 [1/s^2]$ Isosurface at Time $t/T = 0$ (Baseline)

a consequence of this additional structure the pressure side limb of the horseshoe vortex (PS-HV) is more constrained to the hub end wall. At

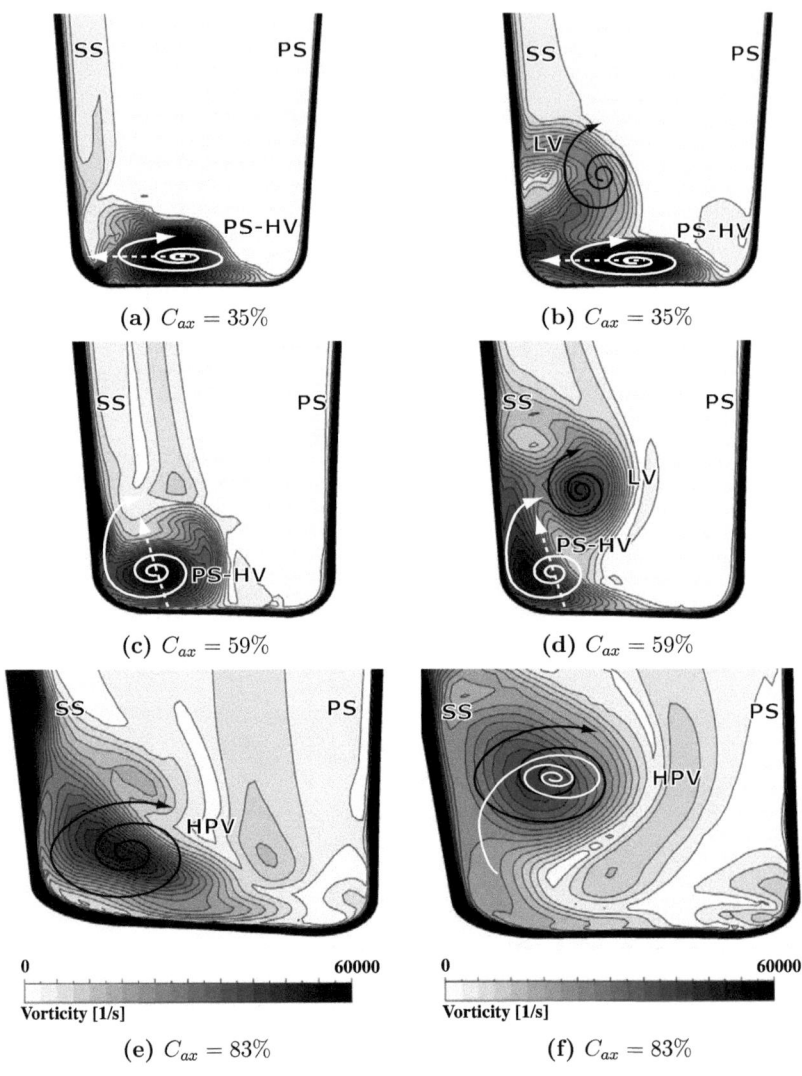

Figure 4.19: Computational Absolute Vorticity in Circumferential Planes in Rotor Passage $t/T = 0$ $IR = -0.1\%$ (left) and $IR = 0.9\%$ (right) (Baseline)

59% axial chord the pressure side limb of the horseshoe vortex (PS-HV) has reached the blade suction side corner. Therefore, the purge flow vortex (LV) is pushed radially outwards. Further downstream the passage vortex starts to migrate radially out. In the purge flow case the two initial vortices merge and become one large diffused structure as seen at 83% axial chord. In the suction case the resulting passage vortex is radially closer to the hub end wall, smaller and stronger in vorticity. If Figures 4.19 (e) and (f) are compared to Figures 4.7 (a) and (b) a similar trend is seen. With purge flow the hub passage vortex is larger, less intense in terms of vorticity and radially further out than in the suction case. The main reason for the different behaviour with purge flow is an additional vortex created by the purge flow jets. This injection vortex has about the same strength as the rotor hub passage vortex.

4.5 Summary

The results with axisymmetric end walls show a total-to-total efficiency drop of $\Delta \eta_{tt} = 0.6\% \pm 0.32\%$ when increasing the injection from the sucking mode $IR = -0.1\%$ to the blowing mode $IR = 0.9\%$. This results in an efficiency sensitivity to purge flow of $-0.6\%/\%$. With injection and conventional end walls the hub secondary flows have a larger radial extent. The vorticity contour plots in the rotor hub passage vortex region show that the vorticity peak values are reduced by using purge flow. However, the integrated circulation of the passage vortex with purge flow has increased by 10%. The circulation of the hub trailing shed vortex has increased by 30% with injection through the rotor upstream rim seal.

From the time-resolved flow field analysis it can be concluded that the rotor hub passage vortex is less stable with purge flow. If the vortex is stretching around the downstream vane leading edge without purge flow the vorticity is increasing to maintain circulation. With purge flow however, the vortex increases in size as a result of this interaction. At the same time the integrated unsteadiness increases. This could be the result of a vortex break down in the purge flow case. Furthermore, the hub pasage vortex center shows much more radial movement dependent on its vane relative position.

The spectral analysis of the time-resolved pressure signal in the hub region shows a subharmonic frequency band around half the blade passing frequency. These frequencies could possibly result from a non-linear combination with some cavity modes. Nevertheless no conclusive statement can be made as there is no time-resolved data from the cavity available.

Based on the time-resolved computation the purge flow mechanism can be summarized as follows. With suction the rim seal exit pressure field is generally an unsteady superposition of the vane and blade induced pressure field. The resulting pressure field in this machine is dominated by the vanes. With purge flow it becomes rotor dominated as a result of a feedback mechanism. The injection jets introduce blockage which causes concave streamline curvature. This results in an elevated pressure level in front of the jets. As a consequence the injection jets are stationary in the relative frame and are modulated by the first vane row. These jets carry normal vorticity as they leave the rim seal. Due to turning around the leading edge streamwise vorticity is created. In a circumferential plane at 35% rotor axial chord an injection vortex is detected which shows the same sense of rotation as the rotor hub passage vortex. This vortex merges with the rotor hub passage vortex. As a result the hub passage vortex becomes larger and less vortical as described in the experimental part.

5 Influence of Purge Flow on End Wall Profiling

In the two previous chapters the flow mechanisms and performance impact of end wall profiling and purge flow have been assessed independently. In this chapter the combined effects of these two mechanisms are addressed. Up to now end walls are essentially designed in absence of purge flow. However, Milli and Shahpar [?] showed in a computational investigation that non-axisymmetric end walls are very sensitive to the effects of secondary cooling mass flows. They numerically demonstrated that most benefits achieved by a design based on a clean annulus model were lost when cooling mass flows were considered. In this chapter the first experimental data of the performance sensitivity of non-axisymmetric end walls to purge flow is presented. Additionally the sensitivity to purge flow over a range of injection rates was investigated for the second end wall design. Finally the time-resolved computations are used once more to assess the behaviour of the purge flow structures inside the blade row.

5.1 Measured Efficiency Sensitivity to Purge

In this section the measured total-to-total efficiency is plotted for all three end wall geometries for suction and purge flow. With the second end wall design two additional injection rates have been measured. The definition of efficiency used in this study, accounting for the injection is given in equation 3.2. Table 5.1 gives the measured absolute efficiencies for the measured purge flow cases. In Figure 5.1 the purge flow efficiency numbers from Table 5.1 are plotted together with the efficiency numbers at small suction $IR = -0.1\%$ as given in Table 3.2. Furthermore the absolute expanded uncertainty of $\pm 0.37\%$ for the total-to-total efficiency is drawn as errorbars in Figure 5.1.

As seen in Figure 5.1 the first end wall design with suction gives an improvement of more than $1\% \pm 0.32\%$ relative to the baseline case with suction.

	$IR = 0.5\%$	$IR = 0.9\%$	$IR = 1.3\%$
Baseline		90.4% ± 0.37%	
1. Gen.		90.8% ± 0.37%	
2. Gen.	90.9% ± 0.37%	90.6% ± 0.37%	90.2% ± 0.37%

Table 5.1: Measured Absolute Efficiency for Purge Flow

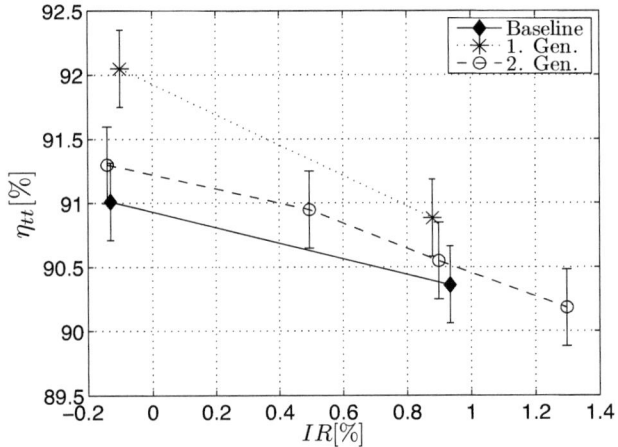

Figure 5.1: Measured Efficiency Response to Injection Purge Flow for Three End Wall Geometries

As mentioned in the chapter 3 this improvement is essentially the result of an improved first nozzle row. However, when purge flow is applied the benefit diminishes to only 0.5%±0.32%. Therefore, the resulting sensitivity is $-1.2\%/\%$. With the baseline geometry the sensitivity as seen in chapter 4 is half as big with $-0.6\%/\%$.

The second end wall design with suction shows an efficiency increase of 0.3% ± 0.32% relative to the baseline case. As discussed in chapter 3 this increase in efficiency is mainly the result of the improved rotor performance compared to the baseline geometry. Because the second non-axisymmetric end wall design for the first nozzle guide vane causes an increase in loss relative to the baseline. With the second end wall design two additional injection rates were measured. The average efficiency sensitivity for the

5.1 Measured Efficiency Sensitivity to Purge

	EXP	CFD
Baseline	−0.6%/%	−0.58%/%
1. Gen.	−1.2%/%	−0.67%/%
2. Gen.	−0.7%/%	−0.65%/%

Table 5.2: Measured Efficiency Sensitivity for Purge Flow

second end wall between $IR = -0.1\%$ and 0.9% is with $-0.7\%\%$ slightly higher than in the baseline case. The efficiency sensitivity curve of the second end wall design shows a kink at $IR = 0.5\%$. In section 5.3 this effect will be discussed in detail. Table 5.2 summarizes the averaged sensitivities for all three end wall geometries between suction $IR = -0.1\%$ and the standard purge flow case at $IR = 0.9\%$. In Table 5.2 additionally the efficiency sensitivities to purge flow of the time-resolved calculations for all end walls are given. These values have been calculated based on the absolute efficiency numbers from the time-resolved calculations as given in Table 5.3. The predicted sensitivity for the baseline and the second end wall design case are matching quite well with the measured sensitivity. However, the strong sensitivity of the first end wall design case is not captured by the computation. The reason for this is a poor prediction of the benefits due to end wall profiling with suction $IR = -0.1\%$. The efficiencies based on the CFD results of the two non-axisymmetric end wall cases don't show the measured improvement as seen in Table 5.3.

	$IR = -0.1\%$	$IR = 0.9\%$
Baseline	90.66%	90.08%
1. Gen.	90.66%	89.99%
2. Gen.	90.69%	90.04%

Table 5.3: Absolute Efficiency Numbers from Time-Resolved CFD Calculation

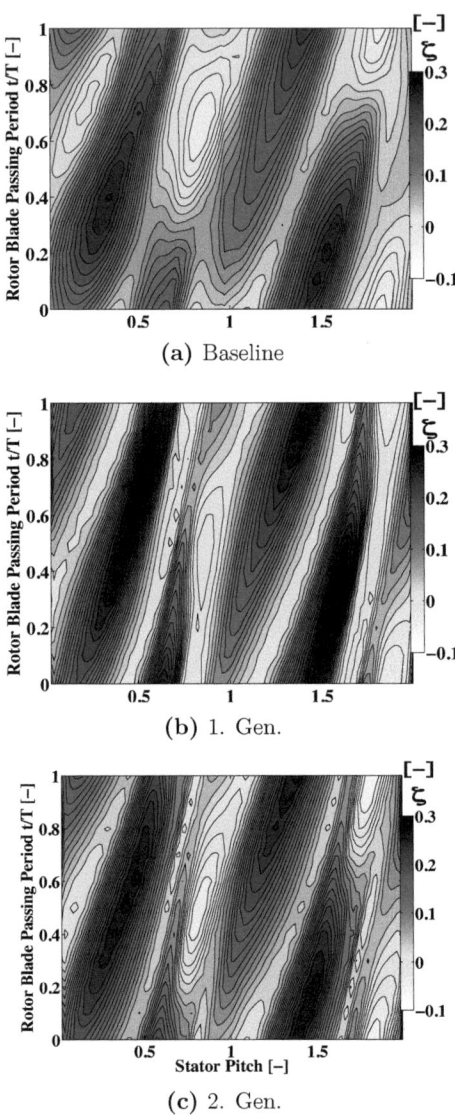

Figure 5.2: Time-Space Diagram of Axially Average Rim Seal Exit Non-Dimesional Static Pressure at $IR = 0.9\%$ from Computation

5.2 Non-Axisymmetric End Walls with $IR = 0.9\%$

5.2.1 Influence of Vane Hub Profiling on Static Pressure at the Rim Seal Exit

As mentioned in the introduction, non-axisymmetric end wall profiling was first introduced to get a more uniform circumferential pressure gradient at the rim seal exit in order to reduce purge flow (Rose [?]). However, the design focus with non-axisymmetric end wall profiling moved towards secondary loss reduction. Doing this the circumferential gradient at the rim has often been increased with end wall profiling. The presented vane hub end wall designs as seen in Figure 3.2 show a dent near the peak suction point. This dent introduces concave curvature on the suction side and therefore, increases the pressure on the suction side. This reduces the cross-pressure gradient. However, this perturbation has to be smoothed out towards the trailing edge in order to have an axisymmetric shape at the rim. As a result, convex streamwise curvature is introduced at the suction side, decreasing the suction side pressure. As a result the circumferential rim pressure gradient is increased in contrast to the initial design intention of Rose [?]. As described in the design methodology section the aim with the second end wall design was to reduce the cross-passage pressure gradient at the vane exit in order to reduce the overturning into the rotor. This was achieved through a 25% smaller amplitude as well as through a ridge at the trailing edge as seen in Figure 3.2. Therefore, the second end wall design shows smaller circumferential vane exit pressure variation. Still the baseline case shows the smallest variation of all three cases.

5.2.2 Non-Dimensional Rim Seal Exit Pressure Field

Figure 5.2 shows again the axially averaged non-dimensional rim-seal exit pressure ζ as defined in Equation 4.2. The quantity is plotted in an absolute frame time-space diagram over two vane pitches and over one blade passing periode. Values above zero are indicative of outflow and vice-versa. Inclined features in the absolute frame time space diagram are associated with the rotor while vertical ones are associated with the first vane row. As already described in section 4.4.1 the rim seal exit flow field based on the pressure field becomes rotor dominated if purge flow is present. The vane blade ratio is two to three. Therefore, the flow field should show three inclined bands of ingestion (bright) and ejection (dark) associated with the three rotor blade rows. However, the non-axisymmetric cases show twice the number of structures as the axisymmetric geometry. Therefore, it can be concluded that a higher harmonic is introduced by the applied profiling. As a result the circumferential gradients become stronger. The first design shows fine bands of intense dark which indicate a strong rim seal ejection. At the same time the second design shows the strongest ingestion. Finally the axisymmetric case shows the strongest modulation over time characterized by stronger temporal (vertical) gradients.

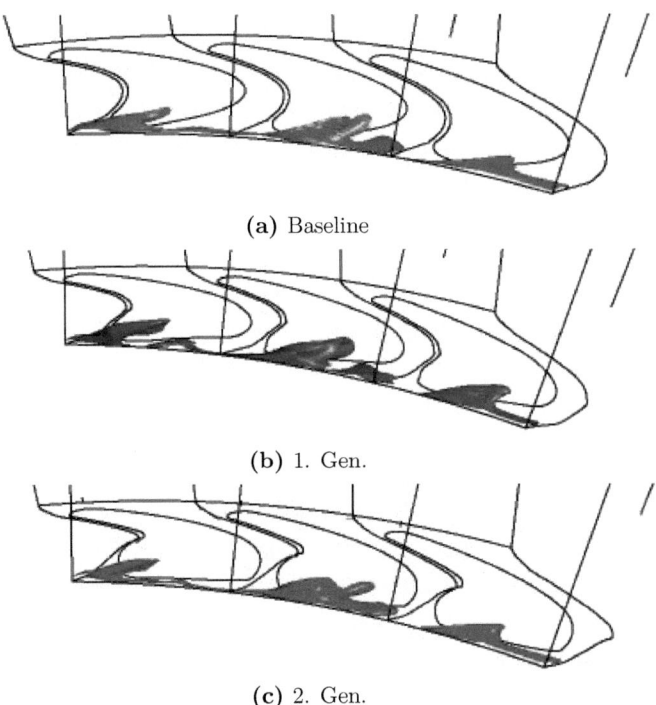

(a) Baseline

(b) 1. Gen.

(c) 2. Gen.

Figure 5.3: Isosurface of Rotary Stagnation Temperature $319K$ at $IR = 0.9\%$ from Computation at Time $t/T = 0$

5.2 Non-Axisymmetric End Walls with $IR = 0.9\%$

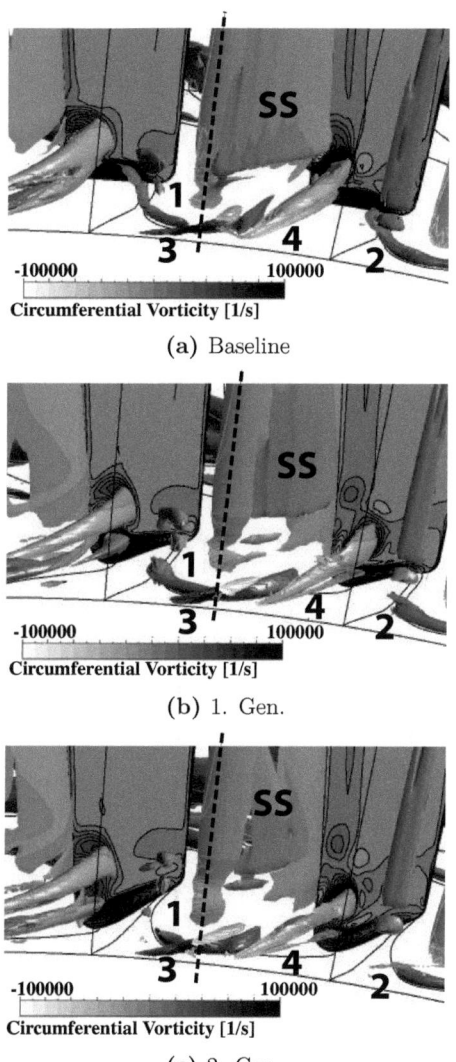

(a) Baseline

(b) 1. Gen.

(c) 2. Gen.

Figure 5.4: Q-Factor $10^7[1/s^2]$ Isosurface at $IR = 0.9\%$ from Computation at Time $t/T = 0$

5.2.3 Purge Flow Structure

In this section the effect of the end wall shape on the development of the purge flow structures is described. As seen in Figure 5.2, the profiling directly influences the rim seal exit pressure field which in turn influences the injection jets. Therefore, Figure 5.3 shows the the rotary stagnation temperature isosurface of $319K$ at $t/T = 0$ for the two non-axisymmetric and the axisymmetric end wall cases at an injection rate of $IR = 0.9\%$ in order to visualize the injection jets. The first end wall design shows the most pronounced jet and the highest penetration in all three passages compared to the other profiled case and the baseline case. This is the consequence of the very confined high ζ regions as seen in Figure 5.2. The second end wall design shows a more circumferential spreading of the injection compared to the other two cases. With the baseline geometry the jets show the strongest variation depending on the vane relative position.

Figure 5.4 shows again the isosurfaces of $Q = 10^7 [1/s^2]$ for the axisymmetric and the two non-axisymmetric end wall cases. The isosurfaces are used to visualize the vortex tubes at blade row inlet. Due to the effective second rotor hub end wall profiling the pressure side leg of the horseshoe vortex pressure side limb **1** is strongly reduced compared to the other two cases as seen in Figure 5.4. With the first end wall profile this structure is hardly reduced compared to the baseline case. The two leakage vortices can again be identified. The anticlockwise injection vortex **4** and the clockwise injection vortex **3**. Judging from Figure 5.3 the jet at this instant is the strongest in the central passage. Therefore, the injection flow structures in this passage are considered. The anticlockwise injection vortex **4** in the central passage is more pronounced with the axisymmetric and the first profiled end wall geometry. The clockwise injection vortex **3** is most pronounced with the first end wall design. Overall the first end wall design shows the larger vortices than the other two cases.

In order to see the effect on streamwise vorticity the axial vorticity in a circumferential plane at 35% rotor axial chord is plotted in Figure 5.5 for all three geometries. In this plane the axial vorticity is close to the streamwise vorticity. The layer of positive vorticity at the hub end wall labeled **1** is the pressure side limb of the horseshoe vortex. The first end wall design reduces the vorticity of this structure slightly compared to the baseline. The second design shows the strongest reduction of this structure. Structures **A − C** show the purge flow induced vortices. With the second end wall design these vortices are weaker than with the baseline or the first end

5.2 Non-Axisymmetric End Walls with $IR = 0.9\%$

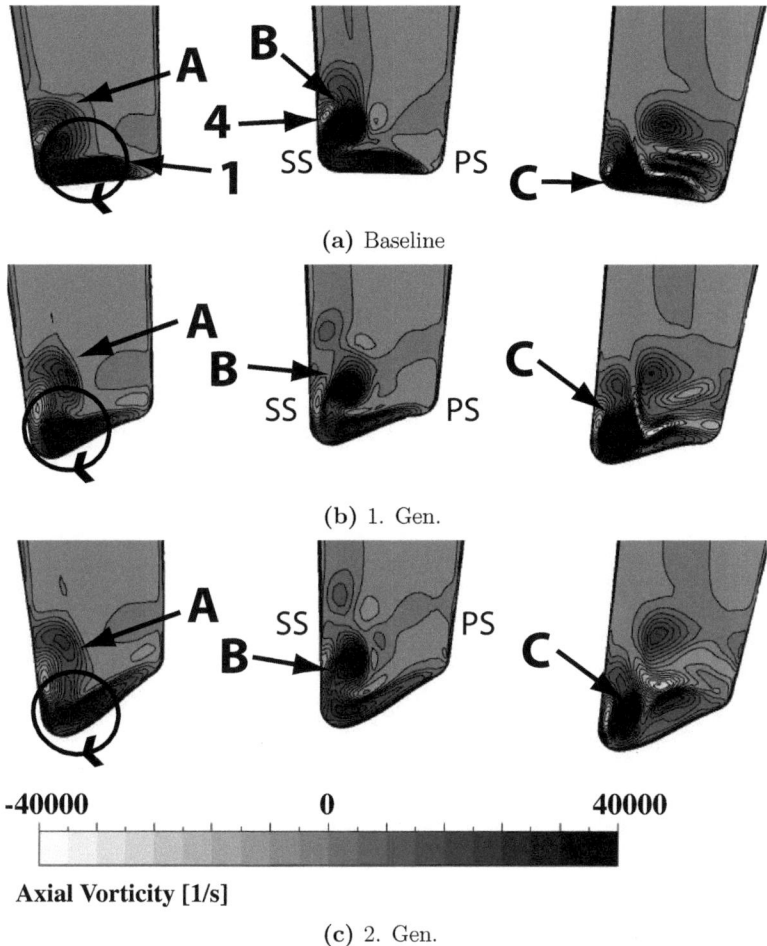

Figure 5.5: Axial Vorticity in an Circumferential Plane at 35% Rotor Axial Chord at $IR = 0.9\%$ from Computation at Time $t/T = 0$

wall design independent of the relative rotor-stator-position. In situation **A** and **C** the leakage vortex with the first end wall design is stronger than the same vortex in the baseline case. In situation **B** they are of about the same strength. The baseline case shows the strongest modulation of the injection vortex as a result of the strongest modulation of the jets seen in Figure 5.3. This is the consequence of the strongest temporal ζ variation as seen in Figure 5.2. The leakage vortex contains about the same level of streamwise circulation as the horseshoe pressure side limb **1** and therefore, about the same effect on the blade row exit hub secondary flow structure. Especially as it will merge with vortex **1** to become one large passage vortex. The structure labeled **4** is the suction side leg of the blade horseshoe vortex. Dependent on the strength of the injection vortex **A − C** this vortex is found in different radial positions. Vortex **4** is moving the most in the baseline case because the injection vortex is varying the most too. The first end wall design creates the strongest suction side leg of the blade horseshoe vortex.

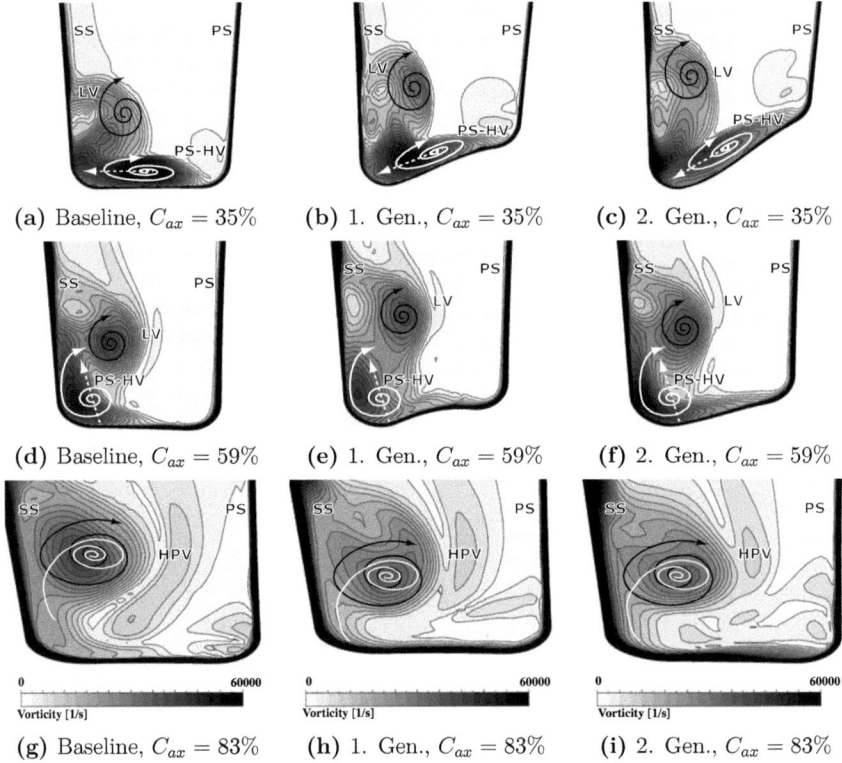

(a) Baseline, $C_{ax} = 35\%$ (b) 1. Gen., $C_{ax} = 35\%$ (c) 2. Gen., $C_{ax} = 35\%$

(d) Baseline, $C_{ax} = 59\%$ (e) 1. Gen., $C_{ax} = 59\%$ (f) 2. Gen., $C_{ax} = 59\%$

(g) Baseline, $C_{ax} = 83\%$ (h) 1. Gen., $C_{ax} = 83\%$ (i) 2. Gen., $C_{ax} = 83\%$

Figure 5.6: Absolute Vorticity in Circumferential Planes in Rotor Passage at $IR = 0.9\%$ from Computation at Time $t/T = 0$

5.2 Non-Axisymmetric End Walls with $IR = 0.9\%$

Figure 5.6 shows the downstream evolution of the purge flow induced vortex (LV) and the blade pressure side limb of the horseshoe vortex. Figure 5.6 shows the absolute vorticity in circumferential planes at three different axial positions which are at 35%, 59% and 83% axial chord (C_{ax}). The baseline geometry on the left side is compared to the first end wall geometry in the middle and the second end wall geometry on the right side. At 35% axial chord the pressure side limb of the horseshoe vortex (PS-HV) is seen as it migrates from the pressure to the suction side. As mentioned before it can be seen that the second end wall profiling is reducing this structure most effectively. The first design shows as well a slight reduction compared to the baseline case. The purge flow induced vortex (LV) is radially further out in the two profiled cases compared to the baseline case. The first end wall design shows the strongest vorticity as seen in Figure 5.5 too. At 59% axial chord the pressure side limb of the horseshoe vortex (PS-HV) has reached the blade suction side corner. As a consequence the purge flow vortex (LV) is pushed radially out. The purge flow induced vortex (LV) is still radially further out in the two profiled cases. However, in terms of strength the difference between the three cases is small. Further downstream the passage vortex starts to migrate radially out and merges with the purge flow induced vortex (LV) and becomes one large diffused passage vortex structure (PV) as seen at 83% axial chord. The resulting passage vortex (PV) in this plane is closer to the end wall with profiling. Furthermore, the baseline case shows the strongest vortex in this location. This means that there has been more dissipation of the secondary flow structures in the two profiled cases compared to the baseline case.

5.2.4 Measured Rotor Exit Flowfield

Figure 5.7 shows the measured streamwise vorticity field at rotor exit. The five vortical structures at rotor exit are drawn into Figure 5.7. **1** is the tip leakage vortex, **2** the tip passage vortex and **4** the tip trailing shed vortex. In contrast to the hub vortices these tip structures are not changed by the injection. **5** is the hub trailing shed vortex and **3** is the hub passage vortex. If the measured vorticity of the passage vortex **3** in Figure 5.7 is compared to the predicted values at 83% axial chord as seen in Figure 5.6 the three end walls show the same behaviour. The baseline case shows 6% more integrated circulation than the first end wall design and 34% more than the second design. The integrated circulation of the hub trailing shed

5 Influence of Purge Flow on End Wall Profiling

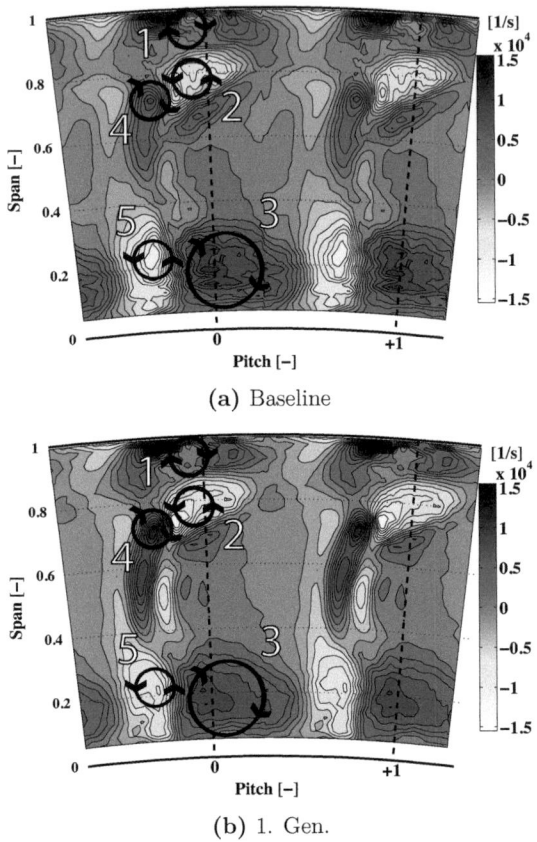

Figure 5.7: Measured Time-Averaged Streamwise Vorticity in the Rotor Frame of Reference at the Rotor Exit at $IR = 0.9\%$

vortex is 13% higher in the baseline case than with the first end wall design. The baseline case shows an even 36% higher value compared to the second design.

The measured **rms** of the total pressure random part at rotor exit is shown in Figure 5.8. The unsteadiness shows a different order concerning the three different end walls than the vorticity does. Therefore, the influence of the end walls on the large scale and small scale structures is not the same. In terms of small scale structures or unsteadiness the first end wall design

5.2 Non-Axisymmetric End Walls with $IR = 0.9\%$

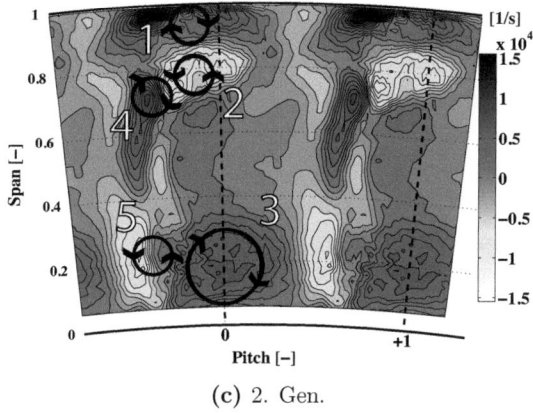

(c) 2. Gen.

Figure 5.7: Measured Time-Averaged Streamwise Vorticity in the Rotor Frame of Reference at the Rotor Exit at $IR = 0.9\%$

shows a very strong hub passage vortex core. If the **rms** distribution of the hub passage and trailing vortex is integrated the first end wall design case shows a 2% higher value than the baseline case. The second end wall design causes 5% lower integrated **rms** than the baseline case. This is the result of more mixing with profiled end walls. The larger structures break down into smaller ones.

5.2.5 Spectral Analysis of Measurements

The frequency composition of the flow in the hub loss core region is determined once more from an FFT of the raw voltage signal of sensor 1 of the FRAP probe, which is directly related to the pressure signal. The amplitude is non-dimensionalized by the baseline value at blade passing frequency. The measurements at 16% span and -12.5% pitch are considered. The blade passing period is $2430Hz$. In Figure 4.12 a comparison of the suction and the purge flow case for the baseline geometry is presented. Figure 5.9 now only shows the frequency content of the three different end wall geometry cases with purge flow. The baseline geometry shows a band of relatively high amplitudes in the subharmonic spectrum of the blade passing frequency with a peak at half blade passing frequency. With the first end wall design this peak is 50% higher and the range of frequencies

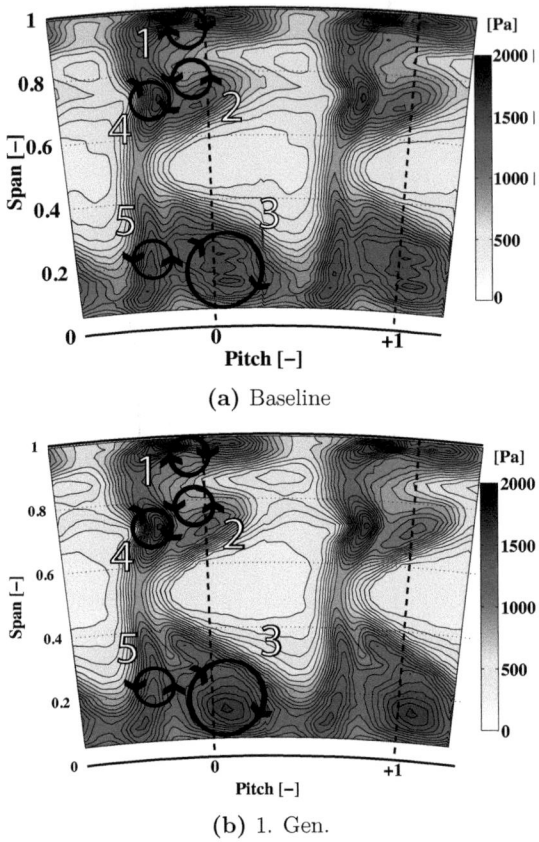

Figure 5.8: Measured Time-Averaged **rms** of the Total Pressure Random Part in the Rotor Frame of Reference at the Rotor Exit at $IR = 0.9\%$

around the peak is much narrower than in the baseline case. This could be the result of some kind of acoustic resonance in the cavity. However, as stated in subsection 4.3.3 no time-resolved cavity data is available. Therefore, no conclusive statement can be made. This potential resonance could be the source of the high **rms** core seen in the hub passage vortex of the first design. With the second end wall shape the blade passing frequency peak is only about 50% of the baseline value. There is also no clear peak in the

5.3 Efficiency Sensitivity Kink

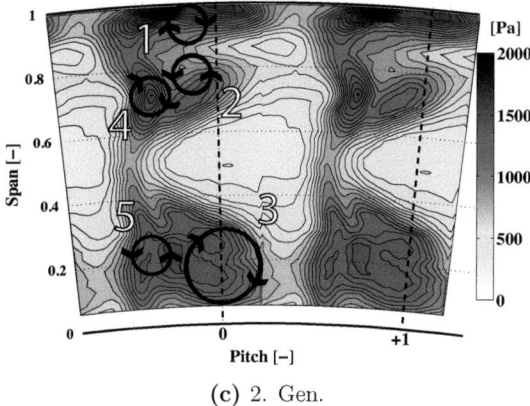

(c) 2. Gen.

Figure 5.8: Measured Time-Averaged **rms** of the Total Pressure Random Part in the Rotor Frame of Reference at the Rotor Exit at $IR = 0.9\%$

subharmonic range. Rather a very broad band of subharmonic frequencies.

5.3 Efficiency Sensitivity Kink

As mentioned in section 5.1 two additional IR-cases have been measured with the second end wall design. One low purge flow case at $IR = 0.5\%$ and one case at the maximal purge flow rate possible with the installed system at $IR = 1.3\%$. The efficiency sensitivity for the second end wall design shows a kink at $IR = 0.5\%$ as seen in Figure 5.1. Below an injection rate of $IR = 0.5\%$ the sensitivity is $-0.6\%/\%$. Beyond this injection rate the sensitivity increases to about $-1\%/\%$.

The power of the turbine drops almost linearly at 2% per percent of injected massflow as seen in Figure 5.10. The change in isentropic work is proportional to the total-to-total pressure ratio change. The total-to-total pressure ratio (plotted in Figure 5.10) is falling as well. As a fact the efficiency is falling by less than one percent. However, the reduction is not linear resulting in the kink seen for the efficiency sensitivity. The mechanism responsible for the reduction of the isentropic work is blockage due to the injected fluid and the losses it causes. The blockage of the rotor by this fluid causes the static pressure to rise at rotor inlet by 0.57% for one

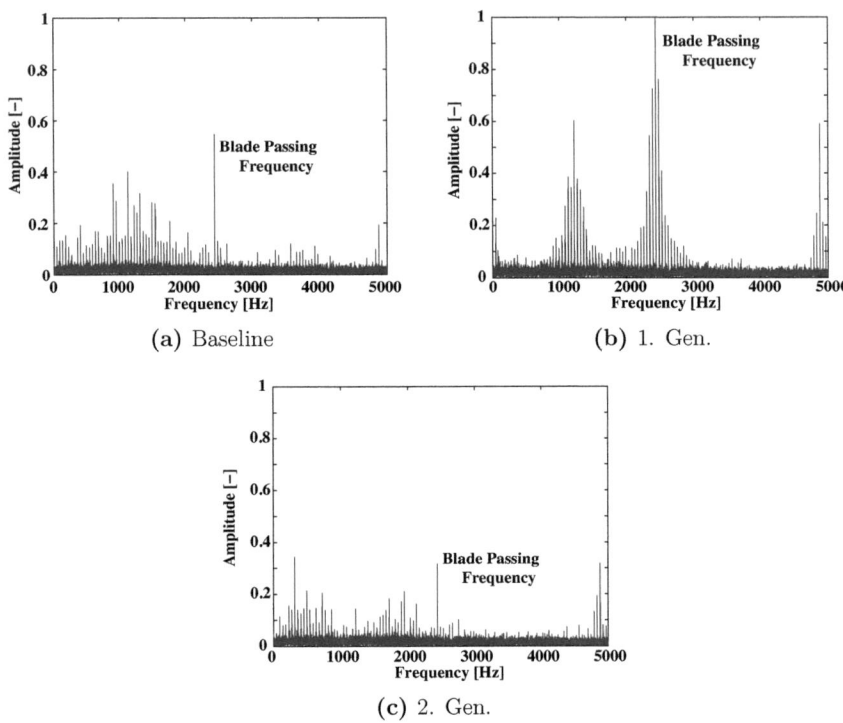

Figure 5.9: Measured FFT of the Raw Voltage Signal of the FRAP Probe at 16% Span and -12.5% Pitch at $IR = 0.9\%$

percent of injected fluid. This does result in a reduction of the first vane exit Mach numbers of -1.2% per injected massflow percent. The effect is not counter-acted by a reduction of static pressure at rotor exit as might be expected, indicating a reduction of capacity in the second vane as well.

5.3.1 Measured Rotor Exit Flowfield

Figure 5.11 shows the circumferentially massflow averaged deviation from the mid-height relative flow yaw angle at the exit of the rotor. The design metal angle is $-67°$ over the whole span. It is apparent that the flowfield at rotor exit is very three-dimensional. At 72% span the tip passage vortex introduces underturning. At 88% span there is the combined overturning of

5.3 Efficiency Sensitivity Kink

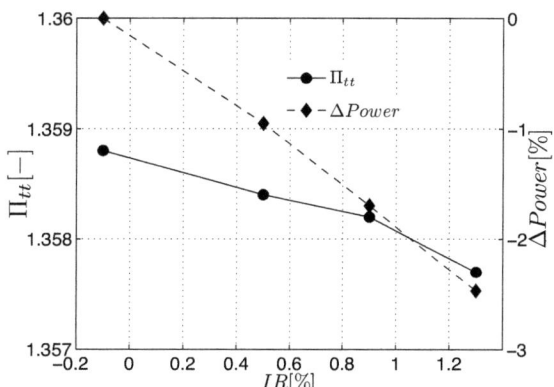

Figure 5.10: Measured Power Drop and Total-to-Total Pressure Ratio with the 2. Design

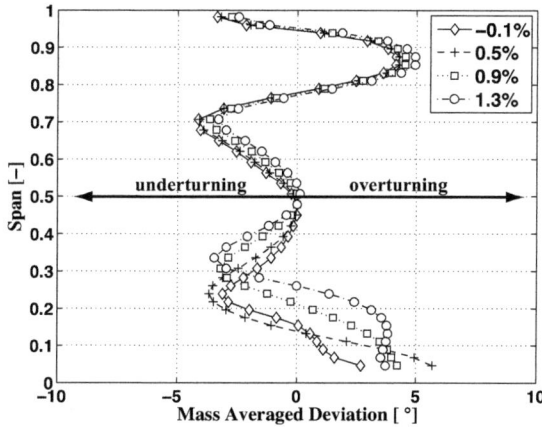

Figure 5.11: Measured Mass-Averaged Deviation at Rotor Exit (2. Gen.)

the tip passage vortex and the tip leakage vortex. Finally at the casing the underturning part of the tip leakage vortex can be seen. The increase in injection rate from $IR = -0.1\%$ to $IR = 1.3\%$ results in a constant increase of $0.5°$ of the flow angle from 50% span to the casing. At the hub one can see the underturning region between 20% and 30% span and the overturning near the hub. At the hub the influence of the injection is very obvious. If the

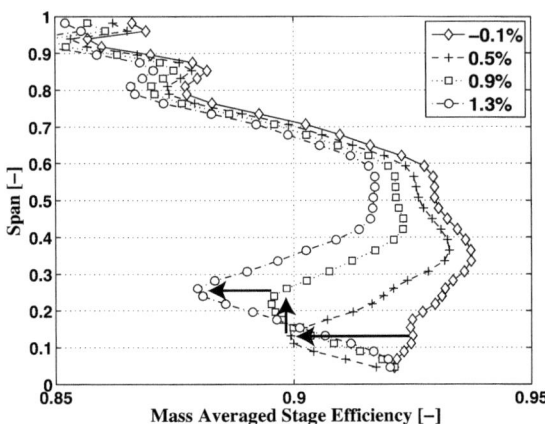

Figure 5.12: Measured Mass-Averaged Total-to-Total Efficiency η_{tt} for Different Injection Rates (2. Gen.)

injection is applied going from $IR = -0.1\%$ to $IR = 0.5\%$ the overturning is increased by $3°$, while the underturning is increased by $0.5°$. When the injection is further increased by 0.4% to an injection rate of $IR = 0.9\%$ the overturning reduces by almost $2°$, while the amount of underturning doesn't change. However, the maximum underturning is now radially further out by about 7% of the span height. At the highest injection rate $IR = 1.3\%$ the overturning reduces by another $0.5°$, while the underturning is increasing by the same amount. The maximal underturning location is 2.3% span further out than with 0.9% injection and the overturning is even 8.6% span further out. The strongest secondary effect can be seen at an injection rate of $IR = 0.5\%$ because this case shows the largest difference between maximum over- and underturning. At an injection rate of $IR = 0.9\%$ the secondary effect reduces before it increases again at an injection rate of $IR = 1.3\%$.

The radial distribution of the total-to-total efficiency is shown in Figure 5.12. It should be noted that this efficiency is linearly related to the total pressure distribution since the power used to calculate it is only a $1D$ value. Therefore, the efficiency in the outer half is linearly decreasing with increasing injection although the total-to-total pressure ratio isn't changing. However, the $1D$ power is dropping linearly as seen in Figure 5.10. There is an increasing efficiency deficit towards the hub for the injection cases.

5.3 Efficiency Sensitivity Kink

With small suction $IR = -0.1\%$ the efficiency plot shows no distinct loss region at the hub as a result of the effective non-axisymmetric end wall profiling. If 0.5% purge flow is used a clear loss region is reappearing. If the purge flow is increased to an injection rate of $IR = 0.9\%$ the size of the loss core as seen as a deficit of efficiency in Figure 5.12 is increasing considerably leading to an increased loss of efficiency per injected massflow as seen in Figure 5.1. At the maximum injection rate of $IR = 1.3\%$ the loss core at the hub is further decreasing with a 2.5% lower minimum. When the injection is increased from $IR = 0.5\%$ to $IR = 0.9\%$ the hub efficiency minimum moves out by 8.6% of the span height. If the injection is increased by the same amount again to $IR = 1.3\%$ the movement is only 4.3% of the span height. As illustrated by the arrows in Figure 5.12 the $IR = 0.9\%$-case shows some transitional behaviour. The efficiency deficit peak isn't rising however, the hub loss core is increasing in size.

Figure 5.13 shows the measured streamwise vorticity at the rotor exit time-averaged in the rotating frame of reference for the four different injection rates. In the outer half of the annulus there is hardly any difference between the four cases therefore, the discussion focuses on the lower half only. In this region two vortical structures of opposite sign can be detected. The larger one with positive vorticity **A** is the rotor hub passage vortex. Just to the left is the trailing shed vortex **B**. If 0.5% purge flow is applied the hub passage vortex center moves 10% of the span height closer to the hub compared to the suction case $IR = -0.1\%$. At the same time the integrated circulation increases by 75%. The integrated circulation of the trailing shed vortex increases by 23%. At the same time the trailing shed vortex center remains at about the same radial position. If the injection rate is increased to $IR = 0.9\%$ the integrated circulation of the hub passage vortex falls by 12% relative to the $IR = 0.5\%$ case. At the same time the hub passage vortex center rises again by 10% span height. The integrated circulation of the trailing shed vortex remains the same as well as the radial position of the vortex center. At the maximum injection rate of $IR = 1.3\%$ the integrated circulation of the passage vortex increases again by 8% and the vortex center is another 10% of the span height further out. The integrated trailing shed vortex circulation increases by 10%, while the center is still at the same radial position.

Figure 5.14 shows the time-averaged **rms** values of the random part of the total pressure signal in the rotating frame of reference. In the suction case $IR = -0.1\%$ there is only one **rms** peak in the region of the hub

5 Influence of Purge Flow on End Wall Profiling

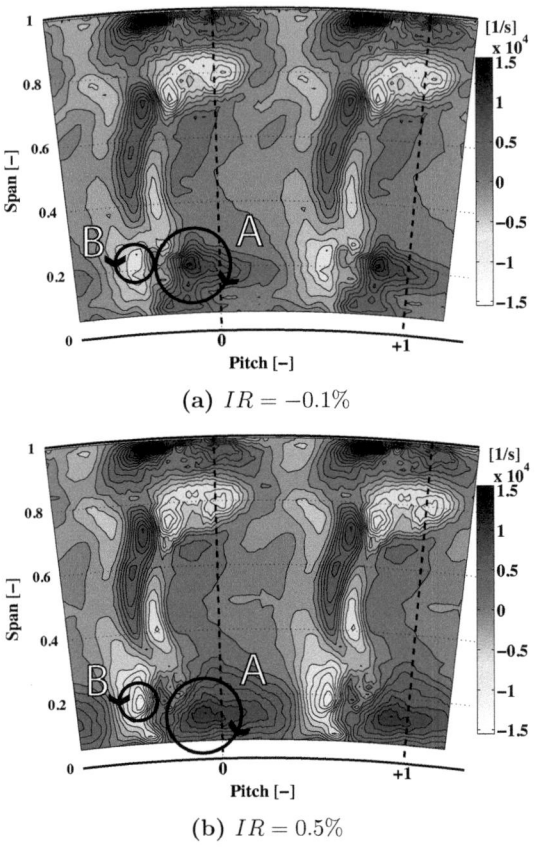

Figure 5.13: Measured Time-Averaged Streamwise Vorticity in the Rotor Frame of Reference at Traverse Plane R1ex (2. Gen.)

trailing shed vortex. The hub passage vortex doesn't show any associated **rms** peak. If 0.5% massflow are injected the integrated **rms** rises by 33% compared to the suction case. If the injection is increased to 0.9% the integrated **rms** value rises by another 16%. The increase is only the result of an increased area because the peak **rms** value remains the same. In the $IR = 1.3\%$ case the hub passage vortex is leaving the end wall and the covered area is slightly decreasing resulting in a 3.5% lower integrated **rms** value. The integrated streamwise circulation and the integrated **rms**

5.3 Efficiency Sensitivity Kink

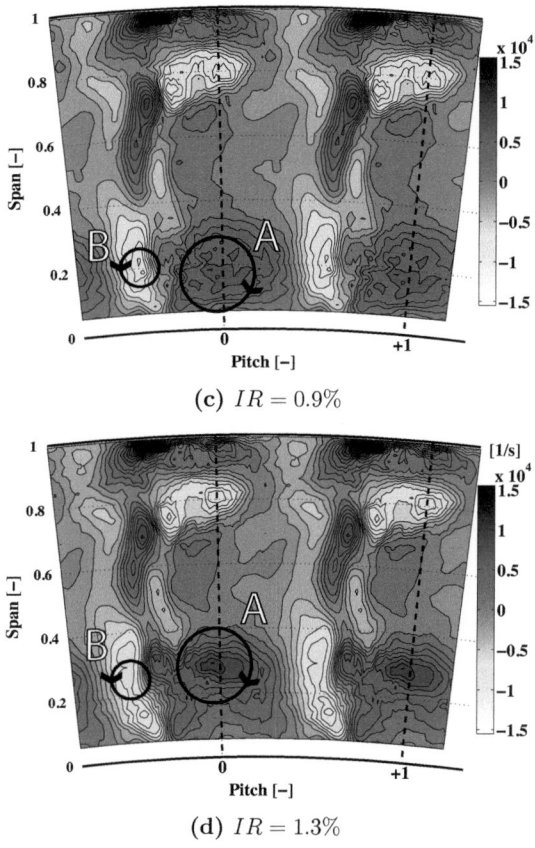

(c) $IR = 0.9\%$

(d) $IR = 1.3\%$

Figure 5.13: Measured Time-Averaged Streamwise Vorticity in the Rotor Frame of Reference at Traverse Plane R1ex (2. Gen.)

of the hub passage vortex show a different development if the injection is increased from $IR = 0.5\%$ to $IR = 0.9\%$. While the **rms** value is rising the circulation value is falling. Therefore, at an injection rate of 0.9% the hub passage vortex is characterized by an increase in small scale features and not by a stronger vortex.

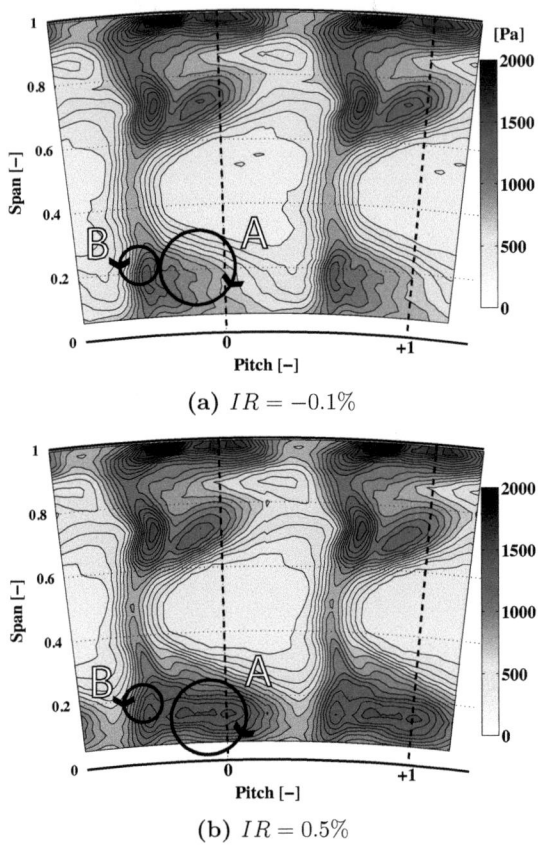

Figure 5.14: Measured Time-Averaged **rms** of Total Pressure Random Part in the Rotor Frame of Reference at Traverse Plane R1ex (2. Gen.)

5.3.2 Spectral Analysis

Figure 5.15 once more shows the frequency composition of the flow in the hub loss core, determined from an FFT of the raw voltage signal of sensor 1 of the FRAP probe. Figure 5.15 shows the frequency content of the three different injection rates for the second end wall design. It can be seen that the already observed subharmonic frequencies are a characteristic of an

5.3 Efficiency Sensitivity Kink

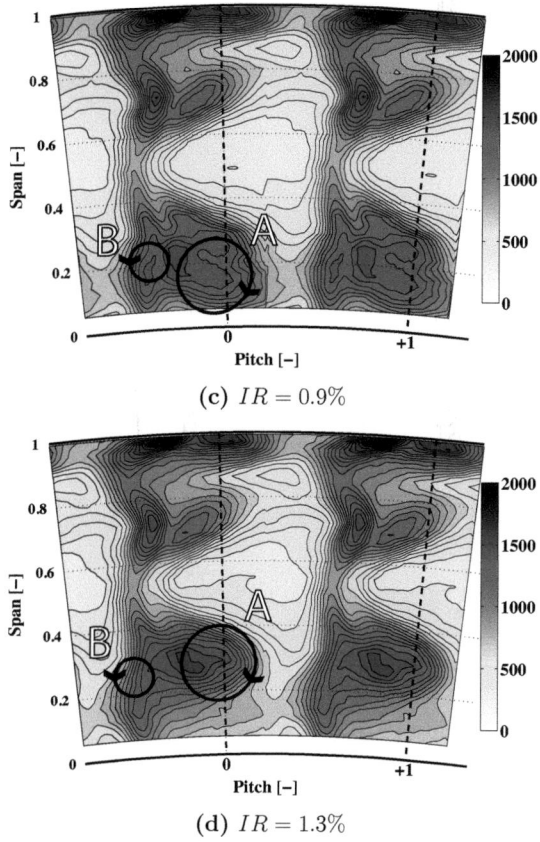

Figure 5.14: Measured Time-Averaged **rms** of Total Pressure Random Part in the Rotor Frame of Reference at Traverse Plane R1ex (2. Gen.)

injection rate of $IR = 0.9\%$ since they are only observed at this particular injection rate.

5.3.3 Pressure Field Sensitivity to Increasing Purge

The kink in the efficiency is the result of a non-linear response of the rim seal exit pressure field to purge flow. Therefore, in Figure 5.16 the axially

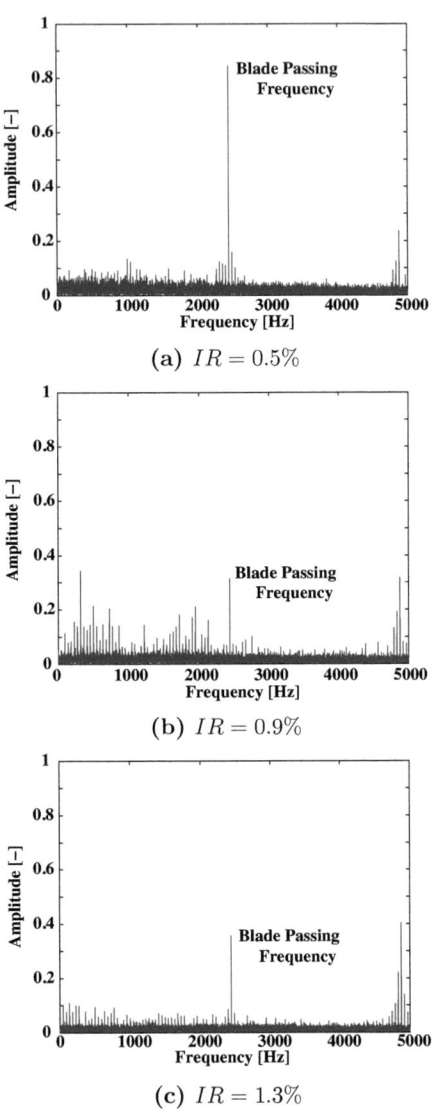

Figure 5.15: Measured FFT at 16% Span and -12.5% Pitch for the Three Injection Rates (2. Gen.)

5.3 Efficiency Sensitivity Kink

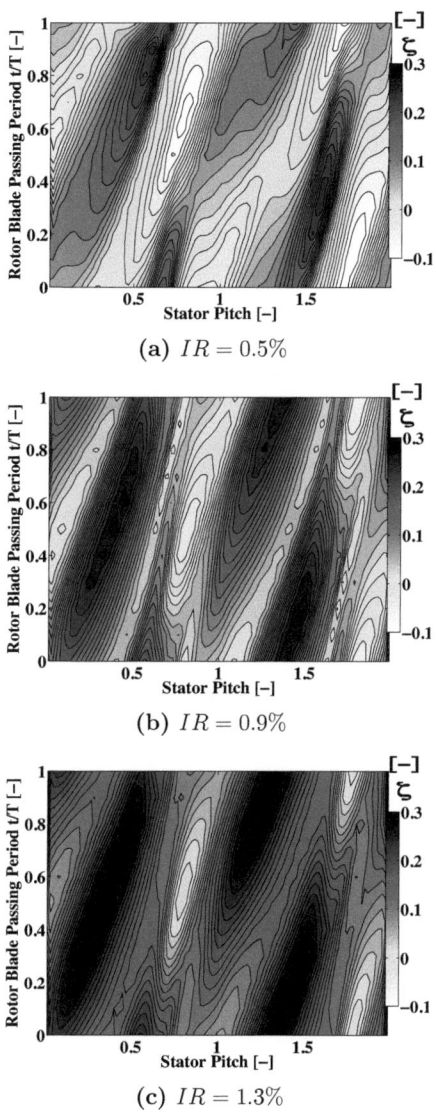

(a) $IR = 0.5\%$

(b) $IR = 0.9\%$

(c) $IR = 1.3\%$

Figure 5.16: Predicted Time-Space Diagram of Non-Dimensional Static Pressure for Three Different Purge Flow Rates

averaged non-dimensional rim seal exit pressure ζ as defined in Equation 4.2 is plotted in an absolute frame time-space diagram for three different purge flow rates. The effect of purge flow reducing the injection can be seen comparing these three pictures. At $IR = 1.3\%$ the ingestion is very small or even absent judging by the pressure field. All three cases show a rotor dominated flow field. Therefore, it can be concluded that the feedback mechanism effect due to blockage of the jets is already of importance at an injection rate of 0.5%. At $IR = 0.5\%$ the modulation of the high pressure region is more profound than in the higher purge flow cases. In the $IR = 0.9\%$ case additionally a second harmonic of the blade passing frequency is appearing in the time-space diagram. At the highest injection rate $IR = 1.3\%$ the influence of the second harmonic is strongly diminished again. If the injection rate is increased the range of the ζ quantity is diminishing. However, the circumferential gradients are the strongest with an injection rate of $IR = 0.9\%$. Generally it can be stated that the pressure field is responding much stronger to an increase in purge flow from 0.5% to 0.9% than to the same increase from 0.9% to 1.3%.

5.3.4 Jet Structure and Vorticity Development with Increasing Purge

The purge flow jets are visualized by the rotary stagnation temperature isosurface at $319K$ as seen in Figure 5.17. Going from 0.5% to 0.9% injection the jets become much more pronounced. Interestingly a further increase of the purge flow rate to $IR = 1.3\%$ leads not to stronger but less pronounced jets. Therefore, a higher injection rate doesn't nescesseraly mean more penetration of the jets.

The pressure field of the $IR = 0.9\%$ case shows the least modulation of the high ejection traces seen as dark regions in Figure 5.16. Therefore, the jets are well pronounced independent of the vane relative position. The second harmonic seen in the $IR = 0.9\%$ case non-dimensional pressure field as seen in Figure 5.16 (b) introduces a second jet as seen in the middle passage of Figure 5.17 (b). The $IR = 0.5\%$ case shows the strongest modulation of the jet caused by the vane pressure field. The first passage as seen in Figure 5.17 (a) shows a small pronounced jet, the middle passage shows something more like a layer and the third passage shows hardly any injection.

Figure 5.18 shows the axial vorticity in the 35% rotor axial chord plane. The three main flow structures are the rotor horseshoe vortex pressure side

5.3 Efficiency Sensitivity Kink

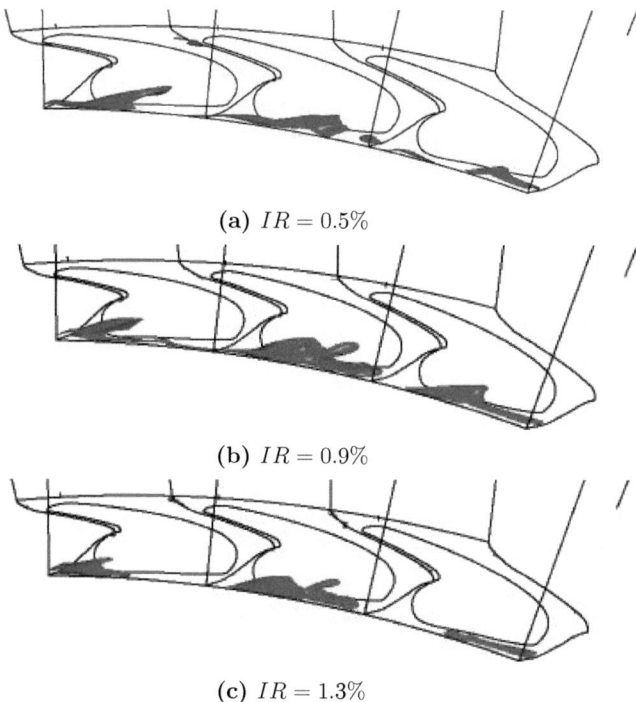

(a) $IR = 0.5\%$

(b) $IR = 0.9\%$

(c) $IR = 1.3\%$

Figure 5.17: Isosurface of Rotary Stagnation Temperature $319K$ (2. Gen.) from Computation at Time $t/T = 0$

limb **1**, the rotor horseshoe vortex suction side limb **4** and the purge flow induced vortex **A** − **C**. All three structures are strongly influenced by a change of the injection rate. Figure 5.18 shows really the effect of the purge flow injection variation in terms of vorticity. The different vorticity levels are the reason for the measured difference seen at blade row exit. The horseshoe vortex pressure side limb is strongly increasing if the purge flow rises from $IR = 0.5\%$ to $IR = 0.9\%$. However, an increase beyond $IR = 0.9\%$ doesn't show a change in the vorticity field. The horseshoe vortex suction side limb **4** is even more influenced as it is trapped between the blade suction side and the injection vortex **A** − **C**. There are two effects associated with the horseshoe vortex suction side limb. Firstly the vortex is diminishing with an increasing injection rate. Secondly as the injection vortex is increasing and radially further out structure **4** is pushed radially

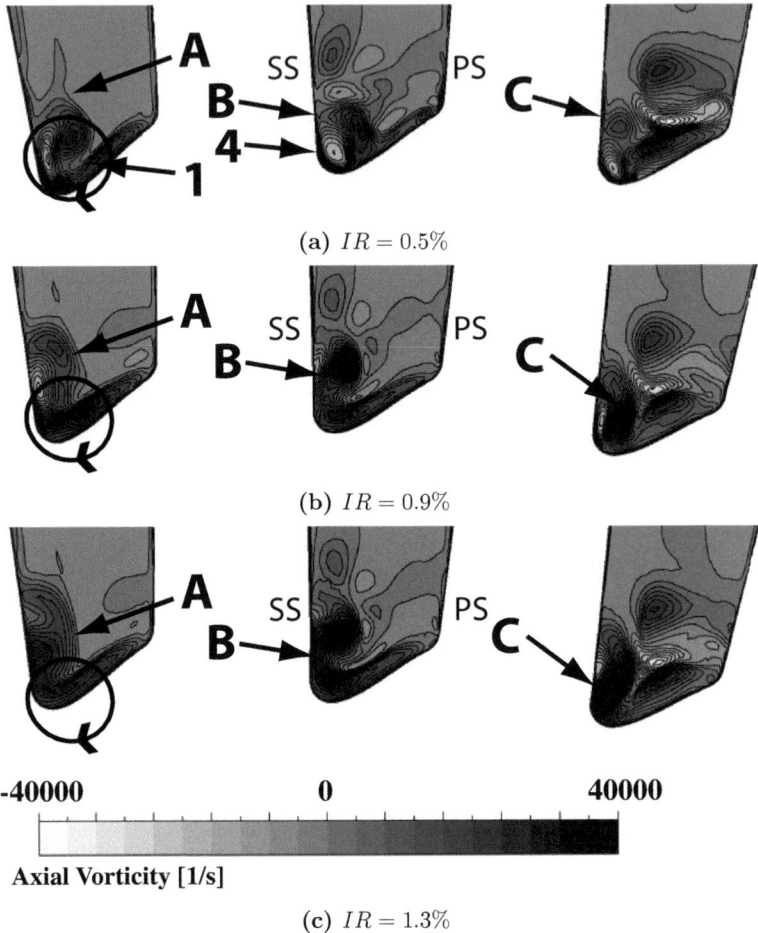

Figure 5.18: Axial Vorticity in a Circumferential Plane at 35% Rotor Axial Chord (2. Gen.) from Computation at Time $t/T = 0$

out as well. At the lowest injection rate $IR = 0.5\%$ the injection vortex is modulated the most by the vane relative position. Therefore, the shape and intensity of the leakage vortex is strong in situation **A** and weak in situation **C**. At the two higher injection rates the vane influence is reduced. Also the injection vortex is radially further out and stronger than at an injection rate of 0.5%. An increase in injection rate beyond $IR = 0.9\%$ does show a minor influence in terms of the purge flow induced vortex. The vorticity is slightly decreasing if the injection rate is increased from $IR = 0.9\%$ to $IR = 1.3\%$. Generally the computed sensitivity of the flowfield to an 0.4% increase of the injection rate from $IR = 0.5\%$ to $IR = 0.9\%$ is stronger than to the same increase from $IR = 0.9\%$ to $IR = 1.3\%$ and agrees with the measured efficiency sensitivity.

5.4 Summary

Three different end wall geometries had their sensitivity to purge flow experimentally assessed. The three geometries are the axisymmetric baseline case and the two non-axisymmetric end wall designs. All three geometries were tested with a small suction setting $IR = -0.1\%$ and the purge flow setting with an injection rate of 0.9% of the main massflow. With the second end wall design two additional measurements at a low injection rate ($IR = 0.5\%$) and at the maximal injection rate ($IR = 1.3\%$) were executed. All three end wall geometries show a deterioration of efficiency as purge flow is added. The baseline turbine shows the least sensitivity with -0.6% $\Delta\eta$ per % injection. The first end wall design shows a efficiency sensitivity for injection of $-1.2\%/\%$ which is twice as big as with the axisymmetric end wall with $-0.6\%/\%$. With the second end wall geometry the sensitivity was reduced to $-0.7\%/\%$. The predicted sensitivity of the baseline geometry and the second end wall geometry do correspond well to the measured sensitivities with $-0.58\%/\%$ and $-0.65\%/\%$ respectively. However, the sensitivities of the first end wall case is underestimated by about a factor 2. The reason is a poor prediction of the benefits of non-axisymmetric end wall profiling with small suction.

The rim seal exit pressure field is a essential part in the purge flow mechanism and the associated efficiency sensitivity. The first end wall design is characterized by strong over-turning at the first vane hub rim which is the result of a stronger stator rim circumferential pressure gradient. If the non-axisymmetric perturbation is smoothed out towards the vane rim an opposite curvature is introduced increasing the cross pressure gradient compared to the axisymmetric end wall. Therefore, the time-resolved rim seal exit pressure shows a strong variation in circumferential direction resulting in very low pressure regions which will cause the strongest ejection jets with the first end wall design. For the second design vane hub end wall it was intended to reduce the strong overturning at the hub. This was mainly achieved by the introduction of a ridge just a the trailing edge. As a result the circumferential pressure gradient is reduced compared to the first design and therefore, the rim seal exit pressure field shows smaller gradients in the circumferential direction. In turn the strength of the injection jets is reduced resulting in a reduced strength of the purge flow induced vortex compared to the first end wall design.

The experimental evidence at rotor exit shows an increase in the non-deterministic rms unsteadiness in the rotor hub region for the first end wall design which is 2% stronger relative to the axisymmetric case. The second end wall design shows even a decrease of 5% relative to the datum case. The rms unsteadiness is believed to be related to turbulent dissipation so an increase implies more mixing and loss. In contrast to the small scale structures (**rms**) the baseline case shows the strongest large scale structure in terms of the highest integrated circulation value due to reduced mixing. The subharmonic frequency content based on a spectral analysis of the experimental data inside the rotor hub passage vortex shows a clear peak at half the blade passing frequency for the first end wall design which could be an acoustic resonance. The other two end wall geometries rather show a broad band of subharmonic frequencies.

Based on the four injection rates tested an efficiency sensitivity curve can be plotted for the second end wall design. This curve shows a kink at an injection rate of $IR = 0.5\%$. Below this injection rate the sensitivity is $-0.6\%/\%$. Beyond this injection rate the sensitivity increases to about $-1\%/\%$. The kink in the efficiency curve is coupled to a change of mode of the rotor hub passage vortex.

The non-dimensional rim seal exit pressure field from the time-resolved calculations shows the strongest variation in circumferential direction at an injection rate of 0.9%. This is mainly the result of the appearance of a second harmonic which introduces as well a second injection jet. Furthermore the $IR = 0.9\%$ case shows the least influence of the stator row. Increasing the injection rate from 0.5% to 0.9% has a much stronger effect in terms of the pressure field, than increasing the injection by the same amount from 0.9% to 1.3%. This effects the introduced vorticity and results finally in the steeper slope of the sensitivity curve.

From the measured data at rotor exit the following effects can be extracted. If the purge flow is increased to $IR = 0.5\%$ the rotor hub passage vortex gains intensity and is closer to the end wall. At $IR = 0.9\%$ the passage

5.4 Summary

vortex is increasing considerably in size and losing intensity at the same time. However, the unsteadiness increases by 16%. At the same time the vortex lifts off the end wall. At $IR = 1.3\%$ the hub passage vortex is becoming a more confined structure again. The intensity of the vortex is increasing resulting in an increase of the integrated circulation by 8%. The unsteadiness falls to the same level as at $IR = 0.5\%$. The spectral analysis of the probe signal shows that the subharmonic frequencies are only present at an injection rate of $IR = 0.9\%$. Therefore, it can be concluded that the subharmonic frequency effect is strongly coupled with a particular range of injection massflow.

6 Conclusion and Contribution

The key findings can be summarized as follows:

- a new approach to calculate stream wise vorticity from a single plane measurement was developed

- a maximal increase of efficiency of more than $1\% \pm 0.32\%$ is achieved with non-axisymmetric end wall profiling

- the main contribution in the first design comes from the first vane row profiling

- with high blockage blading fillet radii have a strong influence on the secondary flow field and should therefore be included in design CFD meshes

- unexpectedly the end wall profiling not only leads to reductions in the secondary flow regions but also at mid height.

- this is due to an alleviation of vane blockage and consequent reduction of Mach numbers and aft loading to delay transition.

- about half of the benefits of the profiling are lost when 1% of purge flow is applied

- therefore, purge flow can not be neglected during the design phase

- purge flow introduces a flow structure with stream wise vorticity which merges with the rotor hub passage vortex leading to a very diffused zone

- with purge flow present end wall profile designing should aim to reduce the circumferential pressure gradient at the rim as much as possible

- measurements at additional purge flow rates for one end wall case show a kink in the efficiency sensitivity which is associated with the lift off of the blade row hub passage vortex

Frozen Flow Approach

This work introduces a new approach to calculate the axial derivatives of the velocities with a single traverse plane and time-resolved data. 'Frozen Flow Assumption' means that a flow structure such as the passage vortex is not changing its shape between two time-steps. This approach is predominately used to calculate stream wise vorticity for which the three dimensional vorticity vector is needed. Essentially this means a halving of the measurement time compared to a multi-plane approach. The approach was validated against a multi plane measurement. Furthermore it is shown that the Helmholtz equation well suited for incompressible cascades isn't an appropriate alternative for rotating machines with compressibility effects.

End Wall Profiling

The first non-axisymmetric design shows an improvement in total-to-total stage efficiency of more than $1\% \pm 0.32\%$ relative to the baseline case efficiency of $91\% \pm 0.37\%$. With the second design an improvement of $0.3\% \pm 0.32\%$ compared to the baseline is achieved.

The first design is characterized by a large improvement in the first vane row. The large improvement is not only the result of suppressed secondary flow but also due to an alleviation of vane blockage and consequent reduction of Mach numbers. Furthermore the strong aft loading caused by the non-axisymmetric vane tip profiling leads to a delayed transition causing a further loss reduction. The time-resolved flow field analysis shows that the improved vane additionally results in a 1% higher freestream relative total pressure ratio. As a consequence these structures will provide more unsteady work compared to the baseline case contributing to an even higher efficiency.

The second design shows two important things. Firstly does the inclusion of the fillets during the design phase result in an improved prediction of the secondary flow field. This leads to the improved second rotor design. The integrated circulation is reduced by about 50% compared to the baseline case leading to a loss reduction twice as big as with the first rotor design. Secondly it is seen that to extreme curvature can potentially lead to a separation as seen in the tip suction side corner. This leads to 9% more loss in the outer annulus half for the second design compared to the baseline case resulting in the worst performance of all three vanes.

Based on the three cases the relationship between measured secondary kinetic energy and loss reduction was examined. The two hub end walls show a linear relationship between loss and secondary kinetic energy reduction. The ratio between loss reduction and secondary kinetic energy reduction is 1 : 5 for the first vane hub end wall. For the rotor hub end wall the ratio is less favorable with 2 : 25. This is because not only secondary losses are considered but the entire loss. At the first vane tip no linear relationship is found because the responsible loss does not originate from secondary flow. Secondary kinetic energy still serves as a good design optimization parameter for the reduction of secondary losses. However, in this study it has been shown that end wall profiling can also significantly change the profile loss by altering the peak velocities as a result of reduced blockage. A further point are changes in the loading profile. In this work it has been shown that aft-loading can have a beneficial effect.

Purge Mechanism and Effect on the Rotor Hub Secondary Flow Field

Based on the axisymmetric case time-resolved computations the purge flow mechanism can be summarized as follows. Without the influence of purge flow ejection the pressure field at rim seal exit is essentially an unsteady superposition of the vane and blade pressure fields. In this turbine it is rather a vane dominated pressure field. However, with purge flow present the pressure field is also influenced by the injection itself resulting in a feedback mechanism. The blockage of the purge flow jets is leading to streamline curvature. As a result the pressure in front of the jets is increased. The injection jets remain circumferentially in approximately the same rotor relative position. Additionally these jets create two counterrotating streamwise normal vortices. As these vortices are turned around the leading edge, streamwise vorticity is created. When these streamwise vortices are merging with the rotor hub passage vortex, a very diffused flow structure is seen at rotor exit which results in a reduction of the streamwise vorticity. However, at the same time the integrated circulation and the integrated unsteadiness in terms of **rms** are increased by 10% and 30% respectively. Finally the higher unsteadiness means also 18% higher loss resulting in a total-to-total efficiency drop of $0.6\% \pm 0.32\%$ per injected massflow percent. All this is accompanied by subharmonic frequencies which are only found with purge flow.

Purge Flow Sensitivity of the Three End Walls

If purge flow is applied the non-axisymmetric end walls lose about half of their benefits. The reason is a higher sensitivity of the non-axisymmetric turbines to purge flow. The first very successful end wall design shows a total-to-total efficiency sensitivity of $-1.2\%/\%$ which is twice as big as with the baseline geometry. The second end wall design shows with $-0.7\%/\%$ a slightly higher sensitivity than the baseline case with $-0.6\%/\%$. The reason are different circumferential pressure gradients at the first vane rim. The first design profiling leads to the strongest circumferential pressure gradient. As a consequence the first end wall design shows the strongest injection jets of all three cases. The measured data at blade row exit shows that this results in a more intense high **rms** zone in the hub secondary flow region. With the second end wall it was the design intention to reduce the overturning at the hub. Therefore, the circumferential pressure gradient at the rim is reduced. The conclusion is that end wall design activity in an environment of secondary cooling flows should not only focus on secondary loss reduction, but also on a reduced circumferential pressure gradient at the rim.

Change of Mode at an Injection Rate of 0.9%

With the second profiled end wall geometry measurements at four injection rates were executed. From this data the following conclusions can be drawn. The total-to-total efficiency curve shows a kink at an injection rate of 0.9%. This kink is associated with the lift off of the hub passage vortex. At 0.9% the vortex becomes unstable an leaves the end wall accompanied by the highest unsteadiness, an increase in area and a decrease in vorticity. Small purge flow (0.5%) leads to an intensification of the rotor hub passage vortex. At the highest injection rate of 1.3% the hub passage vortex is reestablishing at a higher radial position and is again more confined. Furthermore the vorticity increases again and the unsteadiness is falling. As seen from the spectral analysis the subharmonic frequencies are only appearing at the injection rate of 0.9%. Therefore, it can be concluded that these frequencies are most probably the result of a cavity mode strongly coupled with the particular injection rate of 0.9%.

6.1 Suggestion for Further Work

Secondary kinetic energy is still a reliable design target function. However, new quantities have to be included which could capture the effect of the trailing shed vorticity in low aspect ratio turbines and potential separations as observed in this work. Furthermore it is seen that successful profiling is characterized by a certain end wall pressure shape. This knowledge should as well be translated into an additional optimization parameter even more as pressure is very well predicted by CFD codes.

The time-resolved flow field analysis gives some indication for improvement. So far only the free stream relative total pressure ratio μ as a representative quantity has been considered. With the capability of entropy measurements the entropy of the vane flow structures of different vane and blade end walls should be studied extensively.

Considering the combined effect of purge flow and end wall profiling it should be tried to see how the sensitivity of the profiled cases could be reduced. One potential route is the reduction of the circumferential pressure gradient at the vane rim. In order to design end walls including the effect of purge flow efficiently a representative model of the purge flow is needed. Generating such a model the following details have to be considered. First it is a time-resolved mechanism and it is even not a simple superposition of the vane and blade pressure fields. There is also a feedback mechanism of the jets themselves. Furthermore due to inertia of the flow there could be a potential time lag of the jets relative to the pressure field fluctuations at rim seal exit.

Finally it should be investigated if the findings concerning the mode shift at 0.9% injection are of generic nature or if they are a particular characteristic of the investigated second design end wall geometry.

Bibliography

[1] B.J. Abu-Ghannam and R. Shaw. Natural Transition of Boundary Layers - The Effect of Turbulence, Pressure Gradient and Flow History. *Journal of Mechanical Engineering Science*, 22(5):213–228, 1980.

[2] M. J. Atkins. Secondary Losses and End-Wall Profiling in a Turbine Cascade. *IMechE*, C255/87:29–42, 1987.

[3] T. Behr. *Control of Rotor Tip Leakage and Secondary Flow by Casing Air Injection in Unshrouded Axial Turbines*. PhD Thesis, Diss. ETH No. 17283, 2007.

[4] Th. E. Biesinger. *Secondary Flow Reduction Techniques in Linear Turbine Cascades*. PhD Thesis, University of Durham, 1993.

[5] A. Binder, W. Forster, K. Mach, and H. Rogge. Unsteady Flow Interaction Caused by Stator Secondary Vortices in a Turbine Rotor. *Journal of Turbomachinery-Transactions of the ASME*, 109(2):251–257, 1987.

[6] T. C. Booth. Rotor-Tip Leakage Part I - Basic Methodology. *Journal of Engineering for Power*, 104:154–161, 1982.

[7] J. Boudet, N. J. Hills, and J. W. Chew. Numerical Simulation of the Flow Interaction Between Turbine Main Annulus and Disc Cavities. In *ASME Turbo Expo, GT-2006-90307*, Barcelona, 2006.

[8] G. Brennan, N. W. Harvey, M. G. Rose, N. Fomison, and M. D. Taylor. Improving the Efficiency of the Trent 500 HP Turbine Using Non-Axisymmetric End Walls: Part I: Turbine Design. In *ASME Turbo Expo, 2001-GT-0444*, 2001.

[9] V. S. P. Chaluvadi, A. I. Kalfas, and H. P. Hodson. Blade Row Interaction in a High-Pressure Steam Turbine. *Journal of Turbomachinery-Transactions of the ASME*, 125(1):14–24, 2001.

[10] J. W. Chew, S. Dadkhah, and A. B. Turner. Rim Sealing of Rotor-Stator Wheelspaces in the Absence of External Flow. *Journal of Turbomachinery-Transactions of the ASME*, 114(2):433–438, 1992.

[11] N. A. Cumpsty. *Compressor Aerodynamics*. Longman, 1989.

[12] S. Dadkhah, A. B. Turner, and J. W. Chew. Performance of Radial Clearance Rim Seals in Upstream and Downstream Rotor-Stator wheelspaces. *Journal of Turbomachinery-Transactions of the ASME*, 114(2):439–445, 1992.

[13] W. N. Dawes. *A Computer Programm for the Analysis of Three-Dimensional Viscous Compressible Flow in Turbomachinery Blade Rows*. Manual for BT0B3D, 1991.

[14] C. D. Dean. On the Necessity of Unsteady Flow in Fluid Machines. *Journal of Basic Engineering -Transactions of the ASME*, 81(1):24–28, 1959.

[15] M. E. Dejc and A. E. Zarjankin. Methods of Increasing the Efficiency of Turbine Stages. *Teploenergetika*, 2:18–24, 1960.

[16] J. D. Denton. Loss Mechanisms in Turbomachines. *Journal of Turbomachinery-Transactions of the ASME*, 115(4):621–656, 1993.

[17] V. Dossena, A. Perdichizzi, and M. Savini. The Influence of Endwall Contouring on the Performance of a Turbine Nozzle Guide Vane. *Journal of Turbomachinery-Transactions of the ASME*, 121(2):200–208, 1999.

[18] Y. Dubief and F. Delcayre. On Coherent-Vortex Identification in Turbulence. *Journal of Turbulence*, 1(1):11–11(1), 2000.

[19] A. Duden, I. Raab, and L. Fottner. Controlling the Secondary Flow in a Turbine Cascade by 3D Airfoil Design and Endwall Contouring. In *ASME Turbo Expo, 98-GT-072*, 1998.

[20] J. Dunham. A Review of Cascade Data on Secondary Losses in Turbines. *Journal of Mechanical Engineering Science*, 12(1):48–59, 1970.

[21] S. Eymann, W. Foerster, M. Beversdorf, U. Reinmoeller, R. Niehuis, and J. Gier. Improving 3D Flow Characteristics in a Multistage LP

Turbine by Means of Endwall Contouring and Airfoil Design Modification: Part I: Design and Experimental Investigation. In *ASME Turbo Expo, GT-2002-30352*, 2002.

[22] T. Germain, M. Nagel, and R.-D. Baier. Visualization and Quantification of Secondary Flows: Application to Turbine Bladings with 3D-Endwalls. In *Proceedings of the 8th ISAIF, Lyon*, Lyon, 2007.

[23] S. Girgis, E. Vlasic, J.-P. Lavoie, and S. H. Moustapha. The Effect of Secondary Air Injection on the Performance of a Transonic Turbine Stage. In *ASME Turbo Expo, GT-2002-30340*, Amsterdam, 2002.

[24] C. Gossweiler. *Sonden und Messsystem für schnelle aerodynamische Strömungsmessung mit piezoresistiven Druckgebern*. PhD Thesis, Diss. ETH No. 10253, Zuerich, Switzerland, 1993.

[25] D. G. Gregory-Smith, C. P. Graves, and J. A. Walsh. Growth of Secondary Losses and Vorticity in an Axial Turbine Cascade. *Journal of Turbomachinery-Transactions of the ASME*, 110(1):1–8, 1988.

[26] S. Harrison. The Influence of Blade Lean on Turbine Losses. *Journal of Turbomachinery-Transactions of the ASME*, 114(1):184–190, 1992.

[27] J. C. Hartland, D. G. Gregory-Smith, N. W. Harvey, and M. G. Rose. Nonaxisymmetric Turbine End Wall Design: Part II - Experimental Validation. *Journal of Turbomachinery-Transactions of the ASME*, 122(2):286–293, 2000.

[28] N. W. Harvey, D. A. Newman, G. Brennan, and M. G. Rose. Improving Turbine Efficiency Using Non-Axisymetric End Walls: Validation in the Multi-Row Environment and with Low Aspect Ratio Blading. In *ASME Turbo Expo, GT-2002-30337*, Amsterdam, 2002.

[29] N. W. Harvey, M. G. Rose, M. D. Taylor, S. Shahpar, J. C. Hartland, and D. G. Gregory-Smith. Nonaxisymmetric Turbine End Wall Design: Part I - Three-Dimensional Linear Design System. *Journal of Turbomachinery-Transactions of the ASME*, 122(2):278–285, 2000.

[30] W. R. Hawthorne. Secondary Circulation in Fluid Flow. *Proceedings of the Royal Society of London. Series A, Mathematical and Physical Sciences*, 206(1086):374–387, 1951.

[31] H. P. Hodson and W. N. Dawes. On the Interpretation of Measured Profile Losses in Unsteady Wake-Turbine Blade Interaction Studies. *Journal of Turbomachinery-Transactions of the ASME*, 120(2):276–284, 1998.

[32] G. L. Ingram, D. G. Gregory-Smith, M. G. Rose, N. W. Harvey, and G. Brennan. The Effect of End-Wall Profiling on Secondary Flow and Loss Development in a Turbine Cascade. In *ASME Turbo Expo, GT-2002-30339*, 2002.

[33] ISO. *Guide to the Expression of Uncertainty in Measurement (GUM)*. International Organisation for Standardisation (Geneva, Switzerland), ISBN 92-67-1011889, 1st edition, 1993.

[34] R. Jakoby, T. Zierer, L. deVito, K. Lindblad, J. Larsson, D. E. Bohn, J. Funcke, and A. Decker. Numerical Simulation of the Unsteady Flow Field in an Axial Gas Turbine Rim Seal Configuration. In *ASME Turbo Expo, GT2004-53829*, Vienna, 2004.

[35] A. Jameson and T. J. Baker. Multigrid Solutions of the Euler Equations for Aircraft Configurations. In *AIAA Paper No. 84-0093*, 1984.

[36] N. Kobayashi, M. Matsumato, and M. Shizuya. An Experimental Investigation of a Gas-Turbine Disk Cooling System. *Journal of Engineering for Gas Turbines and Power-Transactions of the ASME*, 106(1):136–141, 1984.

[37] P. Kupferschmied. *Zur Methodik zeitaufgelöster Messungen mit Strömungssonden in Verdichtern und Turbinen*. PhD Thesis, Diss. ETH No. 12474, Zuerich, Switzerland, 1998.

[38] L. S. Langston. Secondary Flows in Axial Turbines - A Review. *Heat Transfer in Gas Turbine Systems, Annals of the New York Academy of Siences*, 934(1):11–26, 2001.

[39] L. S. Langston, M. L. Nice, and R. M. Hooper. 3-Dimensional Flow within a Turbine Cascade Passage. *Journal of Engineering for Power-Transactions of the ASME*, 99(1):21–28, 1977.

[40] R. Marini and S. Girgis. The Effect of Blade Leading Edge Platform Shape on Upstream Disk Cavity to Mainstream Flow Interaction of a

High-Pressure Turbine Stage. In *ASME Turbo Expo, GT2007-27429*, Montreal, 2007.

[41] C. McLean, C. Camci, and B. Glezer. Mainstream Aerodynamic Effects Due to Wheelspace Coolant Injection in a High-Pressure Turbine Stage: Part I - Aerodynamic Measurements in the Stationary Frame. *Journal of Turbomachinery-Transactions of the ASME*, 123(4):687–696, 2001.

[42] C. McLean, C. Camci, and B. Glezer. Mainstream Aerodynamic Effects Due to Wheelspace Coolant Injection in a High-Pressure Turbine Stage: Part II - Aerodynamic Measurements in the Rotational Frame. *Journal of Turbomachinery-Transactions of the ASME*, 123(4):697–703, 2001.

[43] R. N. Meyer. The Effect of Wakes on the Transient Pressure and Velocity Distribution in Turbomachines. *Journal of Basic Engineering -Transactions of the ASME*, 80(1):1544–1552, 1958.

[44] A. Milli and S. Shahpar. Full-Parametric Design System to Improve the Stage Efficiency of a High-Fidelity HP Turbine Configuration. In *ASME Turbo Expo, GT2008-51348*, 2008.

[45] B. Mischo. *Axial Turbine Rotor Aero-Thermal Blade Tip Performance Improvement Through Flow Control.* PhD Thesis, Diss. ETH No. 17813, Zuerich, Switzerland, 2008.

[46] A. W. H. Morris and R. G. Hoare. Secondary Loss Measurements in a Cascade of Turbine Blades with Meridional Wall Profiling. In *75-WA/GT-13*, 1975.

[47] S. H. Moustapha, G. J. Paron, and J. H. T. Wade. Secondary Flows in Cascades of Highly Loaded Turbine Blades. *Journal of Engineering for Gas Turbines and Power-Transactions of the ASME*, 107(4):1031–1038, 1985.

[48] M. Nagel, L. Fottner, and R.-D. Baier. Optimization of Three Dimensionally Designed Turbine Blades and Side Walls. In *ISABE-2001-1058*, 2001.

[49] J. H. P. Ong, R. J. Miller, and S. Uchida. The Effect of Coolant Injection on the Endwall Flow of a High Pressure Turbine. In *ASME Turbo Expo, GT2006-91060*, Barcelona, 2006.

[50] G. Paniagua, R. Denos, and S. Almeida. Effect of the Hub Endwall Cavity Flow on the Flow-Field of a Transonic High-Pressure Turbine. *Journal of Turbomachinery-Transactions of the ASME*, 126(4):578–586, 2004.

[51] R. Parker and J. F. Watson. Interaction Effects Between Blade Rows in Turbomachines. *Proc. of I.Mech.E*, 186(21), 1972.

[52] A. Pfau. *Loss Mechanisms in Labyrinth Seals of Shrouded Axial Turbines*. PhD Thesis, Diss. ETH No. 15226, Zuerich, Switzerland, 2003.

[53] L. Porreca. *Aerothermal Optimization of Partially Shrouded Axial Turbines*. PhD Thesis, Diss. ETH No. 17138, Zuerich, Switzerland, 2007.

[54] K. Reid, J. Denton, G. Pullan, E. Curtis, and J. Longley. The Effect of Stator-Rotor Hub Sealing Flow on the Mainstream Aerodynamics of a Turbine. In *ASME Turbo Expo, GT-2006-90838*, Barcelona, 2006.

[55] Rolls-Royce. Market Outlook 2007. www.rolls-royce.com.

[56] M. G. Rose. Non-Axisymetric Endwall Profiling in the HP NGVs of an Axial Flow Gas Turbine. In *ASME Turbo Expo, 94-GT-249*, 1994.

[57] M. G. Rose and N. W. Harvey. Turbomachinery Wakes: Differential Work and Mixing Losses. *Journal of Turbomachinery-Transactions of the ASME*, 122(1):68–77, 2000.

[58] M. G. Rose, N. W. Harvey, P. Seaman, D. A. Newman, and D. McManus. Improving the Efficiency of the Trent 500 HP Turbine Using Non-Axisymetric End Walls: Part II: Experimental Validation. In *ASME Turbo Expo, 2001-GT-0505*, 2001.

[59] R. P. Roy, G. Xu, J. Feng, and S. Kang. Pressure Field and MainStream Gas Ingestion in a Rotor-Stator Disk Cavity. In *ASME Turbo Expo, 2001-GT-0564*, New Orleans, 2001.

[60] H. Sauer, R. Muller, and K. Vogeler. Reduction of Secondary Flow Losses in Turbine Cascades by Leading Edge Modifications at the

Endwall. *Journal of Turbomachinery-Transactions of the ASME*, 123(2):207–213, 2001.

[61] J. Schlienger. *Evolution of Unsteady Secondary Flows in a Multistage Shrouded Axial Turbine*. PhD Thesis, Diss. ETH No. 15230, Zuerich, Switzerland, 2003.

[62] M. Sell, J. Schlienger, A. Pfau, M. Treiber, and R. S. Abhari. The 2-Stage Axial Turbine Test Facility LISA. In *ASME Turbo Expo, 2001-GT-0492*, New Orleans, 2001.

[63] O. P. Sharma and T. L. Butler. Predictions of Endwall Losses and Secondary Flows in Axial Turbine Cascades. *Journal of Turbomachinery-Transactions of the ASME*, 109(2):229–236, 1987.

[64] O. P. Sharma, T. L. Butler, H. D. Joslyn, and R. P. Dring. 3-Dimensional Unsteady-Flow in an Axial-Flow Turbine. *Journal of Propulsion and Power*, 1(1):29–38, 1985.

[65] C. H. Sieverding. Recent Progress in the Understanding of Basic Aspects of Secondary Flows in Turbine Blade Passages. *Journal of Engineering for Gas Turbines and Power-Transactions of the ASME*, 107(2):248–257, 1985.

[66] C. H. Sieverding and P. Van den Bosche. The Use of Coloured Smoke to Visualize Secondary Flows in a Turbine Blade cascade. *Journal of Fluid Mechanics*, 134:85–89, 1983.

[67] C. J. Smith. Airline Operating Costs - The Variations. presented at Managing Airline Operating Costs Conference, Lisbon, December 2004. www.sh-e.com/news_presentations.htm.

[68] P. Spellucci. *DONLP2 User's Guide*. Technical University Darmstadt, FB4, AG8, 1995.

[69] H. B. Squire and K. G. Winter. The Secondary Flow in Cascade of Airfoils in a Nonuniform Stream. *Journal Aeronautical Sciences*, 18(4):271–277, 1951.

[70] W. Sturm, H. Scheugenpflug, and L. Fottner. Performance Improvements of Compressor Cascades by Controlling the Profile and Side-

wall Boundary-Layers. *Journal of Turbomachinery-Transactions of the ASME*, 114(3):477–486, 1992.

[71] H. F. Vogt and M. Zippel. Sekundärströmungen in Turbinengittern mit geraden und gekrümmten Schauffeln; Visualisierung im ebenen Wasserkanal. *Forschung im Ingenieurwesen - Engineering Research*, 62(9):247–253, 2001.

[72] I. A. Waitz. Are there practical solutions to reduce noise and emissions? AIAA/AAAF Aircraft Noise and Emissions Reduction Symposium, Monterey, CA, May 2005.

[73] J. A. Walsh and D. G. Gregory-Smith. The Effect of Inlet Skew on Secondary Flows and Losses in a Turbine Cascade. *IMechE*, C275/87, 1987.

[74] A. Weber. *3D Structured Grids for Multistage Turbomachinery Applications based on G3DMESH 1st revision ed.* Institute of Propulsion Technology, German Aerospace Center, Cologne, February, DLR IB-325-05-04, 2004.

[75] D. C. Wisler. The Technical and Economic Relevance of Understanding Blade Row Interaction Effects in Turbomachinery. Blade Row Interference Effects in Axial Turbomachinery Stages, VKI LS 1998-02, 1998.

[76] A. Yamamoto. Production and Development of Secondary Flows and Losses in Two Types of Straight Turbine Cascades: Part I - A Stator Case. *Journal of Turbomachinery-Transactions of the ASME*, 109(2):186–193, 1987.

[77] A. Yamamoto. Production and Development of Secondary Flows and Losses in Two Types of Straight Turbine Cascades: Part II - A Rotor Case. *Journal of Turbomachinery-Transactions of the ASME*, 109(2):194–200, 1987.

A Nomenclature

A.1 Symbols

a	half width of rectangular uncertainty function	$[-]$
C	velocity	$[m/s]$
c	sensitivty coefficient	$[-]$
C_{ax}	axial chord	$[-]$
C_d	discharge coefficient	$[-]$
c_p	specific heat	$[J/kg]$
D	venturi inlet diameter	$[m]$
d	venturi throat diameter	$[m]$
$\overline{f}$	reduced frequency	$[-]$
$\overline{f}$	averaging function	$[-]$
H	humidity	$[\%]$
h	specific enthalpy	$[J/kg]$
IR	injection rate	$[\%]$
K	calibration sensitivity coefficients	$[-]$
k	coefficients of calibration polynom	$[-]$
k	uncertainty coverage factor	$[-]$
M	torque	$[Nm]$
$\dot{m}$	mass flow	$[kg/s]$
N	rotational speed	$[r.p.s]$
p	pressure	$[Pa]$
$\overline{p}$	time averaged pressure part	$[Pa]$
p'	random pressure part	$[Pa]$
$\tilde{p}$	periodic pressure part	$[Pa]$
PR_s	static pressure ratio	$[p_s/p_{t,in}]$
PR_t	total pressure ratio	$[p_t/p_{t,in}]$
PR_trel	relative total pressure ratio	$[p_{trel}/p_{t,in}]$
Q	Q-factor	$[1/s^2]$

R	gas constant	$[J/(kg \cdot K)]$
r	radial coordinate	$[m]$
s	entropy	$[J/(kg \cdot K)]$
S^2	strain rate	$[1/s^2]$
SKE	secondary kinetic energy	$[m^2/s^2]$
T	temperature	$[K]$
T	blade passing periode	$[s]$
t	time	$[s]$
U	blade speed	$[m/s]$
U	voltage	$[V]$
U	blade speed	$[m/s]$
x	axial coordinate	$[m]$
Y	total pressure loss coefficient	$\left[\frac{p_{t,in}-p_{t,ex}}{p_{t,in}-p_{ex}}\right]$
$y+$	non-dimensional wall distance	$[-]$

A.2 Greek

β	venturi diameter ratio	$[-]$
γ	ratio of specific heats	$[-]$
γ	pitch angle	$[°]$
δ	displacement thicknes	$[°]$
ϵ	expansion factor	$[°]$
ζ	non-dimensional rim seal pressure	$[-]$
η	efficiency	$[-]$
θ	circumferential coordinate	$[rad]$
Π	pressure ratio	$[-]$
ρ	density	$[kg/m^3]$
φ	yaw angle	$[°]$
ω	vorticity	$[1/s]$
ω	rotational speed	$[rad/s]$

A.3 Subscripts

atm	atmospheric
by	bypass
$conv$	convection
c, max	compressor maximum
$dist$	disturbance

A.3 Subscripts

dr	drum
ex	row exit
i	velocity component index
in	turbine inlet
low	mid-height to hub
$main$	main flow quantity
n	streamwise normal
$norm$	normalized
$purge$	purge flow quantity
Δr	radial window
r	radial coordinate
ref	reference quantiy
rel	relative frame
$R1ex$	exit blade row
s	static flow quantity
s	streamwise
$S1ex$	exit first nozzle guide vane row
t	stagnation flow quantity
$trel$	relative stagnation flow quantity
$trot$	rotary stagnation flow quantity
tt	total-to-total
$t, f1$	freestreamfluid at rotor inlet
$t, f2$	freestream fluid at rotor exit
$t, w1$	secondary or wake fluid at rotor inlet
$t, w2$	secondary or wake fluid at rotor exit
up	mid-height to tip
x	axial coordinate
v	venturi
γ	pitch angle
θ	circumferential coordinate
φ	yaw angle
1	row inlet
1.5	total-to-static 1.5 stages
2	row exit
∞	freestream

A.4 Abbreviation

B	bypass flow
CFD	computational fluid dynamics
DC	direct current
EXP	measured data
FFT	fast fourrier transformation
FRAP	fast response aerodynamic probe
GUM	guide of uncertainty in measurements
HPV	hub passage vortex
LDA	laser doppler anemometry
LV	purge flow induced vortex
P	purge flow
PIV	paricle image velocimetry
PS-HV	pressure side limb horse shoe vortex
PT100	platinum resistance thermometer
$R1ex$	blade row exit
$R1$	blade row
S	secondary flow
SFC	specific fuel consumption
$S1ex$	first nozzle guide vane exit
$S1$	first nozzle guide vane
$S2$	second nozzle guide vane
$S2ex$	first nozzle guide vane exit
TPV	tip passage vortex
Tr	traverse number
1D	one-dimenional
1. Gen.	first end wall design
2. Gen.	second end wall design
3D	three-dimenional

Die VDM Verlagsservicegesellschaft sucht für wissenschaftliche Verlage abgeschlossene und herausragende

Dissertationen, Habilitationen, Diplomarbeiten, Master Theses, Magisterarbeiten usw.

für die kostenlose Publikation als Fachbuch.

Sie verfügen über eine Arbeit, die hohen inhaltlichen und formalen Ansprüchen genügt, und haben Interesse an einer honorarvergüteten Publikation?

Dann senden Sie bitte erste Informationen über sich und Ihre Arbeit per Email an *info@vdm-vsg.de*.

Sie erhalten kurzfristig unser Feedback!

VDM Verlagsservicegesellschaft mbH
Dudweiler Landstr. 99   Telefon +49 681 3720 174
D - 66123 Saarbrücken   Fax   +49 681 3720 1749
www.vdm-vsg.de

Die VDM Verlagsservicegesellschaft mbH vertritt

Printed by Books on Demand GmbH, Norderstedt / Germany